Proceedings of the First

Science Conference

2013

A NOTE FROM THE EDITOR

Being a scientist is more than about just knowing facts. It is about learning skills and using them to excel in your field. The horizon of human understanding only expands when an individual sees further than their peers and predecessors. Science is a competitive environment for the individual and the most successful will be those that have mastered the skills of being a scientist.

To enable our students to stand under the spotlight of competition the science department at Wreake Valley Academy endeavoured to create a science conference. The event would draw upon the participation of every student studying Applied Science, Biology, Chemisty and Physics A level subjects. The quality of the presentations was consistently high and choosing a winner was particularly difficult. However based on the quality of the presentation, the confidence displayed in its delivery and the prepared way in which questions were responded to the winner was declared to be

Sheena Singadia

The editor would like to thank the following teachers for their work in making the conference a success. Miss Barbi, Dr Billington, Mr Foster, Mr Hill, Dr Kyle-Ferguson, Dr Meakin, Mr Nakeshree, Mr Pabla, Mr Rawson-Gill, Mr Rowles and Mrs Simpson.

Finally the editor would like to thank each and every one of the participants. Your hard work has really paid off and your teachers are very proud of what you have achieved.

Dr David Boyce

CAN PLEASURE RESPONSES FROM CHOCOLATE BE SEEN IN THE EYES

AGGARWAL V.

ABSTRACT

This research is about chocolate and how people's responses can be seen in their eyes. For many years, everyone thought that only the brain was responsible for the response. However, recent evidence suggests that the retina of the eye secretes a substance which the brain normally does when the person eats something sweet. This amazing discovery happened in Philadelphia, at Drexel University. The doctor who came up with it was called Dr Jennifer Nasser.

1. HYPOTHESIS

The hypothesis was very simple. This was that Dopamine was released in the retina of the eyes when the person had something sweet. (Nasser J. , 2013) As soon as the piece of cake would be placed in the mouth, electrical impulses would rise in the retina and Dopamine would be secreted. They also thought that the brain is not the only organ responsible for this substance release. However, there must be something that has to secrete it. But what? The other hypothesis was that neurotransmitters released Dopamine at synapses when exposed to light exposure. This exposure could be sugary food which makes a person happy.

2. TEST

Once this hypothesis was made, Dr Nasser got 9 people from a random sample and made them have a piece of chocolate brownie. They were not allowed to eat it, but the shocking discovery from this test was that the Hypothesis came out true! As soon as the brownie was placed in the mouth of the volunteer, electrical impulses in the retina spiked very high. (Nasser J. D., 2013) This process was called Electroretinography.

3. RESULT

While they did this experiment, the participants were asked to complete a Gormally Binge Eating Scale. This was a questionnaire which assesses the effects of their binge eating behaviour. From this test, (not the questionnaire) they found that Dopamine was released from the eye when the brownie was put in the mouth. They then realised that the exact same Dopamine was released when the participants took an oral drug stimulant called methylphenidate (MPH). This was a stimulant which made them happy. The results from the questionnaire told the scientist that people felt happy when they were exposed to a sugary substance. This then caused Dopamine to be released in the eye.

4. CONCLUSION

To conclude, Dr Nasser found out that Electroretinography could provide a cheap methodology to potentially assess Dopamine responses when exposed to oral stimulation. She also realised that Dopamine would only be released in the retina if the effects of the oral substance made the person excited or happy, as MPH had this effect, and so did the chocolate brownie.

REFERENCES

Nasser, J. (2013, June Monday). Retrieved June Thursday, 2013, from Science Daily: http://www.sciencedaily.com/releases/2013/06/130624111014.htm

Nasser, J. D. (2013). Pleasure response from chocolate. *Obesity*, 976-980.

CARBON DIOXIDE CAN NOW BE TURNED INTO METHANOL, WITH EFFICIENT PROCESSES!

AGGARWAL V.

ABSTRACT

For many years, scientists have tried to reverse the combustion of alcohols, in terms of converting the by-products back in to the alcohol they came from. However, to do this, high temperatures were needed and the costs were too high to the company. Scientists have also come up with catalysts which may lower the activation energy, but never worked out, therefore failing in this experiment. Now, after many years, Professor Frédéric-Georges Fontaine has come up with one combined catalyst, which can make CO_2 be turned back into methanol, with lower greenhouse gases in the process.

1. HYPOTHESIS

There was a very simple hypothesis. One catalyst can be formed from two chemical groups from energy-efficient processes. This will allow the reverse reaction to be made possible, with a very low emission of greenhouse gases. But where will these chemicals come from? Well, molecules found in everyday chemistry are the solution to this hypothesis. (Frédéric-Georges, 2013)

2. TEST

Fontaine captured Borane, which contained Boron, Hydrogen and Carbon. He also got Phosphine- a compound containing Phosphorous, Hydrogen and Carbon. He reacted these two together, but hadn't thought of a name as yet. Once they were at this stage, they then realised that to make the conversion of CO2 into methanol possible required a source of hydrogen and chemical energy. To solve this, they used HydroBorane. The catalyst formed from Borane and Phosphine was used in the reaction, along with HydroBorane. From this, they saw that the hypothesis was correct.

RESULT

From this test, they shown that the reaction was two times more effective than the previous conversion and only a little amount of waste was produced. Because of this, they reduced the amount of greenhouse gases produced and released into the environment. Also, the reaction didn't damage the catalyst, so it was able to be reused again and again.

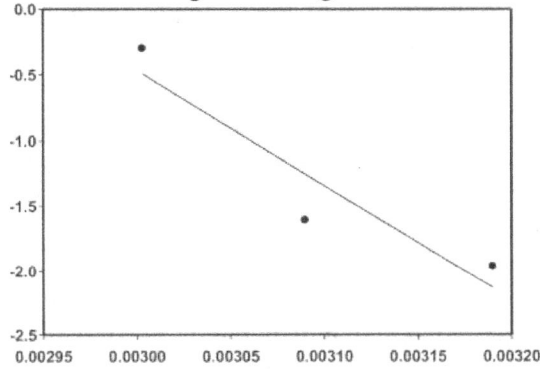

A graph to show the rate of greenhouse gases against the conversion of carbon dioxide into methanol

3. CONCLUSION

To conclude, this means that methanol can be made from the products of the forward combustion reaction, at a lower cost and cheaper efficient processes. This can then be further developed to be fitted into car exhausts so the toxic gas produced can be recycled straightaway.

REFERENCES

Frédéric-Georges, F. (2013, June 20). *Converting Carbon Dioxide into Methanol*. Retrieved June Wednesday, 2013, from Science Daily: http://www.sciencedaily.com/releases/2013/06/130620111230.htm

CONTACT LENSES CAN BE MADE FOR PEOPLE WITH AMD

AGGARWAL V.

ABSTRACT

I researched a very fascinating topic. For many years, people with Age-Related Macular Degeneration (AMD) cannot wear contact lenses as their retina is damaged by this blindness. This means that patients cannot see everything, even with contacts. Scientists have always come up with ways to produce a temporary cure, but have never actually come up with a solution. However, there seems to be a breakthrough in a discovery that allows people with AMD to wear a contact lens that will help them see.

1. HYPOTHESIS

The Hypothesis was very simple. A state of the heart telescopic contact lens can switch between magnified and unmagnified vision for people with AMD. This will allow patients to see clearly and lead a normal life. (J., 2013)

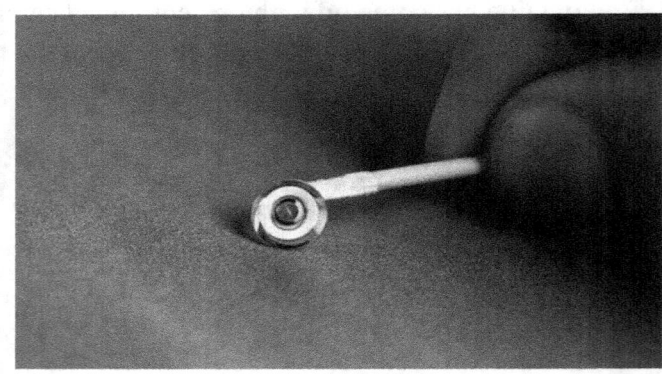

2. TEST

To test this hypothesis, the Professor made a life sized model eye, to capture pictures through the new lens. They used PMMA (polymethyl methacrylate) as the main material. This material is used in most contact lenses. They needed this because tiny grooves were needed to correct the len's shape which could have adjusted the shape of the actual human eye.

RESULT

They found that the magnified image quality through the contact lens was clear and provided a very large field of view. This happened because the lens used tightly fitted mirrors which acted like a telescope. It was inserted into the contact lens witch was just over 1mm thick. The centre of the lens (pupil) provided unmagnified vision, whereas the retina around the outside magnified the view by 2.8x. They then found out that liquid 3D crystal glasses would need to be worn so the view can be switched from magnified and normal. The glasses block the magnifying portion of the lens or the centre which is unmagnified. This is done by polarization.

3. CONCLUSION

To conclude, this meant that the hypothesis was proved right and that people with AMD can see more clearly with this contact lens until a permanent cure is not found for AMD.

REFERENCES

J., F. (2013, June 27). *Contact lens- a telescope for your eye*. Retrieved June 29, 2013, from Science daily: http://www.sciencedaily.com/releases/2013/06/130627125329.htm

ALTURNATIVES TO PLATIMUN CATALYISTS IN HYDROGEN FEUL CELLS
CHRISTIAN JAMES ATTEWELL

ABSTRACT:
Previous research into the area had already found graphene to work as an alternative catalyst at the anode and cathode of a fuel cell, however it was less effective as a catalyst then platinum so to enhance the catalytic properties of graphene the samples were broken down with a ball milling machine so bonds could be made between the halogens (chlorine, bromine, iodine) and graphene, to observe if this enhances catalytic properties of the graphene. The materials were then tested, the material is solution processable, showed imporved electro-catalytic activities needed to facilitate the reactions taking place in a hydrogen fuel cell and A long-term cycle stability.

HYPOTHESIS:
Reacting the halogens (iodine, chlorine and bromine) with graphene, will enhance oxygen reduction reaction properties and create an effective and more viable alternative catalyst to platinum, for the anode and cathode of hydrogen fuel cells.

TEST:
The materials where synthesized by ball-milling the graphite in a planetary ball-mill capsule in the presence of (chlorine, iodine, bromine), respectively. 5.0 g of graphite was placed into a stainless steel capsule containing stainless steel balls (500 g, diameter 5 mm). The capsule was then sealed and degassed. Thereafter, the chosen gas was charged through a gas inlet with cylinder pressure of 8.75 atm. The capsule was then fixed in the planetary ball-mill machine, and rotated at 500 rpm for 48 h. The resultant product where extracted with methanol to get rid of small molar mass organic impurities. HCl solution to remove metallic impurities, if any. Final product was then freeze-dried at −120°C under a reduced pressure (0.05 mmHg) for 48 h to yield 6.09 g for chlorine. 6.93 for bromine and 6.86 g for iodine. The properties of these substances then exerted further test of their properties.

RESULTS: All materials created where found to have improved catalytic properties facilitating oxygen reduction reaction compared to graphene alone, with iodine-edged nanoplatelets performing best, they then found these substances created to be more versatile then platinum (performance reduction did not occur in the presence of carbon monoxide or methonol) whilst they were also more durable maintaining 20% more current then platinum after 10000 cycles.

CONCLUTION:
This was found to be a effective catylist that can revilutionise the future of the hydrogen cell.

REFRANCES:
In-Yup Jeon, Hyun-Jung Choi, Min Choi, Jeong-Min Seo, Sun-Min Jung.
Facile, scalable synthesis of edge-halogenated graphene nanoplatelets as efficient metal-free eletrocatalysts for oxygen reduction reaction.

Article number: 1810, Scientific Reports 3. Research by the universities of South Korea, Case Western Reserve University and University of North Texas from http://www.nature.com/srep/2013/130605/srep01810/full/srep01810.html1518.htm

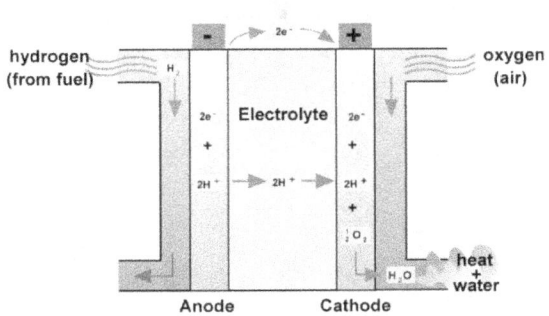

THE USE OF CRYOTHERAPY IN TREATING SOFT TISSUE INJURIES

BROWN. A

ABSTRACT

Cryotherapy is the use of low temperatures to help the healing process of soft tissue injuries. The low temperatures reduce the swelling and pain from any injuries. Research was conducted by Bleakley et al into the effectiveness of cryotherapy. Cryotherapy is the use of low temperatures to help the healing process of soft tissue injuries. The low temperatures reduce the swelling and pain from any injuries.

1. HYPOTHESIS

The test I researched was the use of cryotherapy in the treatment of sports injuries. The hypothesis of the test was that the cold temperatures used during cryotherapy will aid in the healing of soft tissue injuries. Cryotherapy in sport is used to aid in the healing process of injuries sustained during physical exercise.

2. TEST

The test was conducted on 22 patients suffering from different soft tissue injuries. They were given cryotherapy treatment either on its own or in conjunction with other treatments. Two independent reviewers assessed the validity of the results using the PEDro scale

3. RESULT

The results showed that 22 patients met the inclusion criteria of the Physiotherapy Evidence Database (PEDro) scale. There was some evidence that exercise as well as cryotherapy is most effective when recovering from ankle sprains.

4. CONCLUSION

To conclude, this procedure needs more research and more evidence to prove the overall effectiveness of the study

REFERENCES

1. Chris Bleakley, BSc (Hons), MCSP
2. Suzanne McDonough, PhD, MCSP
3. Domhnall MacAuley, MD, FISM

LEISURE ACTIVITIES AND THE RISK OF DEMENTIA

BROWN.H.

ABSTRACT

The risk of dementia increases with increasing age, and preventions of this memory disorder are a health priority. Although participation in leisure activities had been associated with a reduced risk of dementia in later years, no one had actually studied the association in detail, and which leisure activities actually offered the most protection from dementia.

HYPOTHESIS

The Albert Einstein College of Medicine in New York aimed to study the relationship between different leisure activities and the risk of developing dementia in later years, funded by the Bronx Aging Study.

TEST

Using 469 subjects all older than 75, the relationship between leisure activities and dementia was studied. Subjects were screened to see if dementia was present at the baseline. During study visits, neuropsychological tests such as the Blessed Information–Memory–Concentration test, Wechsler Adult Intelligence Scale and the Fuld Object-Memory Evaluation test were used to diagnose dementia by assessing damage to the brain due to an impaired skill. Subjects were formally interviewed about their participation in six cognitive activities (reading, writing, crosswords, group discussions, board games and playing instruments) and eleven physical activities (tennis, swimming, biking, dancing, group exercise, team games, walking, climbing stairs, housework, babysitting and golf). During the study, participation in these activities was verified by family members each time a study visit occurred, allowing the experimenters to add a scale of activities days per week to each subject.

RESULTS

After a study of 5.1 years, dementia had developed in 124 subjects, all of various types such as Alzheimer's disease, vascular dementia and mixed dementia. Out of the leisure activities, only a few were associated with a decreased risk of dementia. Cognitive activities of reading and playing puzzles reduced the risk by 35% and 47%. Only one physical activity showed a reduction, which was dancing (reduced the risk by 76%). These all reduce the risk of dementia as they create new neural pathways in the brain, which usually weakens with age. Creating new neural pathways with these activities will keep the brain stimulated, therefore reducing the risk of dementia. Dancing, as the most effective activity offering protection, would exclude routine dances such as ballroom, whereas taking part in dances with varying steps instead of similar ones are recommended for reducing the risk of dementia.

CONCLUSION

The participation in certain leisure activities is associated with a reduced risk of dementia.

REFERENCES

Fabrigoule C et al, Katzman R, Aronson M- Development of dementing illnesses in an 80-year-old volunteer cohort.

Fuld P et al- Object-memory evaluation for prospective detection of dementia in normal functioning elderly

SOFT NANOFLUIDIC TRANSPORT IN A SOAP FILM

BRUCE C

ABSTRACT

Scientists have found a way to artificially keep fluids moving through the film of a bubble which in turn keeps it floating without popping for much longer. By putting a current through the bubble, the particles of water have more movement so do not succumb to gravity and pop. It has been found that we can change the thickness of the film, and this research could be used to create a new type of micro diode, making parts of a computer smaller, cheaper and faster (al., 2013).

1. HYPOTHESIS

During the test, scientists hoped to find that when passing a voltage through an elastic channel, in this case a bubble, that the fluid between the film would move around stopping the bubble from popping. They expected that as the voltage was increased, the rate at which this fluid moved, the electro-osmotic flow rate would increase at a linear rate.

2. TEST

The scientists placed two electrodes half a centimetre apart. They made a soap film containing water, potassium chloride and a surfactant. The water and ions from the potassium chloride were able to move freely through the film. A voltage was put across the two electrodes, and was increased to see how the change in voltage would affect the movement of the fluid (Yirka, 2013).

3. RESULT

From this experiment, the scientists learnt that when the voltage was put across the electrodes the bubble did in fact not pop as the fluid was able to move. They were surprised however to learn that the electro-osmotic flow rate did not increase linearly. As the voltage was put across, the film of the bubble had thickened allowing for more and faster movement of the fluid, and so it increased at a much faster than linear rate.

4. CONCLUSION

This research has useful applications in computing as it can be used to create a micro-diode, a small device which can be used to let current flow only one way, which will be smaller, faster and cheaper than current alternatives.

REFERENCES

al., B. O. (2013). Soft Nanofluidic Transport in a Soap Film. *Phys. Rev. Lett. 110, 054502*, 5.

Yirka, B. (2013, February 7). *Researchers find soap film micro-channel size tunable with electric charge.* Retrieved June 26, 2013, from phys.org: http://phys.org/news/2013-02-soap-micro-channel-size-tunable-electric.html

CLOUD BEHAVIOUR ON PLANETS IN THE HABITABLE ZONE

BRUCE O.

ABSTRACT

For a planet to be habitable it needs to orbit its star, whilst still maintaining liquid water on it's surface. The formula currently used for calculating the habitable zone largely neglects clouds and any effect they may have on the temperature and climate of the planet. This new research aims to correct this so that a more accurate habitable zone can be calculated.

1. HYPOTHESIS

Cloud behaviour on a planet orbiting a Dwarf Star, could determine whether it is habitable or not, which would lead to the habitable zone being bigger than it is currently considered to be. (Yang, Cowan, & Abbot, 2013)

2. TEST

To test the theory, a 3 Dimensional computer simulation was used to study how air and moisture moves on a tidally locked planet, orbiting a dwarf star. (Koppes, 2013) Any results can be verified once the James Webb Space Telescope has been launched in 2018. This will monitor temperature at different points in a planets orbit.

3. RESULT

The 3D simulations used showed that if there was any water on the surface of the planet, clouds of water vapour would result. They also demonstrate that cloud behaviour has a major cooling effect on the inner portion of the habitable zone. (Koppes, 2013) This means that planets are able to maintain liquid water on their surface much closer to their star than was previously thought possible. This could lead to the habitable zone being bigger than it was previously believed to be.

The results expected from the James Webb Space Telescope are: High temperatures on the 'night side' of the planet, and low temperatures on the 'day side'. These results have been recorded by Earth-observing satellites already. If Brazil is monitored by an Infrared telescope from space, it appears cold, this is due to the fact that it is recording the temperature of the cloud deck, as opposed to the surface temperature, as the water vapour in the clouds absorb infrared radiation, preventing it from reaching the surface.

4. CONCLUSION

This research suggests that there could be double the amount of habitable planets than was previously thought. This could mean that there are 60 billion habitable planets in the Milky Way galaxy alone.

REFERENCES

Koppes, S. (2013, July 1). *CLoud modeling expands estimate of life-supporting planets.* Retrieved July 6, 2013, from UChicagoNews: http://news.uchicago.edu/article/2013/07/01/cloud-modeling-expands-estimate-life-supporting-planets

Yang, J., Cowan, N. B., & Abbot, D. S. (2013). Stabilizing Cloud Feedback Doubles Frequency of Red Dwarf Habitable Planets. *Astrophysical Journal Letters, Vol. 771, No. 2* .

HOW THE NOISE OF BOATS EFFECTS BOTTLENOSE DOLPHIN ECHOLOCATION

CHALLENGER. B

ABSTRACT

Cetaceans navigate and forage acoustically. Anthropogenic noise can spread across long distances underwater; there has been increasing concern that the noise coming from boat engines are causing signal masking of these acoustic behaviours. For this research I have been looking at how the noise of the boats in a harbour has been affecting the behaviour of the Bottlenose Dolphin. Specifically the frequency of echolocation used whilst they are within the perimeter of said boats (Hz).

1. HYPOTHESIS

The new hypothesis is that the bottlenose dolphin ability to echolocate is being affected by the motors for boats; this is due to the frequency at which they would originally perform echolocation is not being used when boats are around. They use a different frequency; one that would normally be used in busier waters. (Elizabeth, 2010)

2. TEST

For the test Elizabeth Shaw along with Cardigan Bay marine wildlife centre recorded results in the summer of 2010. This data included the position of the boat, speed, and distance from dolphin. Also, it included results on the behaviour of the dolphin e.g. frequency of echolocation.

Sound recordings were also made; this was done by lowering a hydrophone into the water when within 300 metres of the dolphin. The hydrophone acts as a pressure transducer; converting the acoustic signals into electronic waves that can be viewed on a spectrogram. The boat engine was turned off during this.

RESULT

The results indicate that the bottlenose dolphins of Cardigan Bay alter the frequency of echolocation clicks when boats are present.

The mean peak frequency of echolocation clicks was significantly higher when no boats were present than when boats were present.

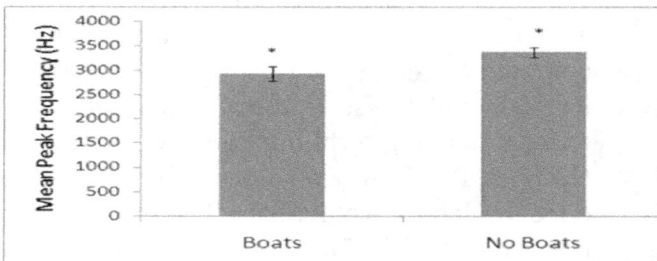

3. CONCLUSION

This indicates that it is the boat noise is what is altering the frequency which the dolphins are using.

REFERENCES

E.Shaw,
AU, W. W. L. 2004. Echolocation signals of wild dolphins. Acoustical Physics, 50, 454-462.

DO MORE MUSCULAR YOUNG MEN LIVE LONGER?

COLEMAN H

ABSTRACT

The objectives of the study were to explore the extent to which muscular strength in adolescence is associated with all cause and cause specific premature mortality (<55 years). It was a prospective, cohort study, set in Sweden. The study consisted of 1 142 599 Swedish male adolescents aged 16-19 years who were followed over a period of 24 years. Of the original 1 142 599 participants, 26, 145 died, the highest forms of mortality found that could be associated with lower muscular strength were suicide, and cardiovascular disease.

1. HYPOTHESIS

Do more muscular young men live longer, due to the fact that having more muscle will give the adolescent boys a better body image, thus avoiding developing mental illnesses such as depression, which could in turn lead to suicide, or due to the fact having for muscle mass indicates a healthier lifestyle, and thus less chance of developing cardiovascular disease. (BMJ, 2012)

2. TEST

The study followed 1,142,599 Swedish male adolescents, aged 16-19 years, who were eligible for military conscription, over a period of 24 years. The study took baseline measurements of height, weight, BMI and blood pressure. The strength measurements were taken from knee extension strength, elbow extension strength and grip strength. After the measurements were taken, they were left alone until death or conclusion of the study.

3. RESULT

During the follow up period of 24 years, 26,145 participants died. Of that number, 1,254 (5.5%) deaths were caused by coronary heart disease, 526 (2.3%) by stroke, 3,425 (14.9%) by any type of cancer, 5,100 (22.3%) by suicide, 5,921 (25.9%) by non-intentional accidents. The remaining deaths, 6,657 (29.1%), were categorised as "other causes of mortality". In summary, greater muscular strength in adolescents was found to be significantly associated with lower risk of premature death from any cause.

4. CONCLUSION

This large cohort study of over a million male Swedish adolescents found associations between greater muscular strength in adolescence and reduced risk of premature mortality from any cause as well as specifically from cardiovascular disease and suicide. There was no association found for deaths from cancer.

REFERENCES

(Choices, 2012) (Ortego, 2012)
(Silventoinen, 2012) (Per Tynelius, 2012)
(Rassmussen, 2012) (BMJ, 2012)

NAKED MOLE RAT IMMUNITY TO CANCER

COOPER D P.

ABSTRACT

Throughout the years of the observation of Naked Mole Rats (Heterocephalus glaber) it has been noticed that they have an extraordinary longevity compared to other rodents of similar size to them; it had also been noticed that a naked mole rat has never developed any type of cancer. This intrigued researchers at the University of Rochester in New York and they carried out several tests/investigations to find out why this is.

1. HYPOTHESIS

To find out the reason why Naked Mole Rats have never developed cancer and thus find a cure for cancer.

2. TEST

The test that was carried out by Vera Gorbunova and her colleagues; they first attempted to grow a type of connective tissue, fibroblasts, in a culture. They noticed that during this growing period, the culture medium became very viscous. The researchers realised that due to them growing fibroblast then this 'viscous' substance must have been the extracellular matrix, which acts as a support for the cells and provides as an internal lubricant. This then started to make more sense as the mole rat lives most of its life burrowing through narrow tunnels and fitting into very tight spaces. Research tne began on this extracellular matrix and it was found that the chemical which was most abundant in it was a polysaccharide called hyaluronan. This chemical is present in the mole rats in a unique form, a heavy weight version, call hich-molecular mass hyaluronan (HMM-HA). Work then began on manipulating the pathways that lead to the build up of HMM-HA in cells to see if cancer growths could be grown in cells which had not been exposed to HMM-MA and comparing it to cells which have.

3. RESULT

The results from this test were that cells that had been exposed to HMM-MA could not develop cancer (*New Scientist*) thus proving that HMM-MA can stop rapid cell growth/ cell growth.

4. CONCLUSION

To conclude, this research shows that HMM-MA can stop cancer and is a cure for it. However, it has not been confirmed that it can stop the development of cancer in humans, and if it is, all of the cells in the body must be manipulated so it can produce or at least come into contact with HMM-MA. This would be very impractical and potentially very dangerous to carry out this procedure. But it is said that it can be used to combat aging of organs and the body and also help alleviate the symptoms of arthritis.

REFERENCES

New Scienctist (2013, June 19). *Naked mole rats reveal why they are immune to cancer.* Retrieved 2013, from http://www.newscientist.com/article/mg21829224.800-naked-mole-rats-reveal-why-they-are-immune-to-cancer.html#.Udmz6fmsjLQ

THE DEVELOPMENT OF AN ALZHEIMER'S PREVENTION DRUG

COOPER D P.

ABSTRACT

My presentation and research was based on the development of a drug that would prevent Alzheimer's disease and help diminish the symptoms of the disease. The disease is thought of as one of the most common causes of dementia in elderly people, and so the development of this drug in an increasing aging population would come in very well. It would also solve the economic, social and emotional burden of the disease.

1. HYPOTHESIS

To find a potential cure/ drug that can stop the development of Alzheimer's once it has been diagnosed.

2. TEST

The tests in which the potential drugs were tested out in were the standard clinical and preclinical drug testing trials. This entails the preclinical test in which drug candidates were tested on growth cultures and also on animals with similar anatomy of humans. This is so we can see if the drug actually has any effect on the disease and that it is safe to use on animals and is not toxic/ has toxic side effects. Then, once it has been deemed safe for usage it will under go three different tests on humans to determine several things.

During phase 1, a small number of healthy volunteers with the disease and all have very similar symptoms will take the drug and the aims is to: asses any human side effects, calculate a safe dosage range, how safe the drug actually is and how well the body uses the drug (like how it is absorbed and expelled from the body).

Phase 2 is carried out on a larger audience with similar symptoms from the disease as side effects that were not witnessed in phase 1 may be observed. This increase in test subjects leads to an even more accurate dosage range and chance to see how well the drug actually does work on stopping the development of Alzheimer's.

In phase 3, the volunteers are normally multinational and have a range of symptoms from the disease but nonetheless have it in some form. This phase is to prove how safe the drug actually is and to monitor side effects t see if they can be minimised. It is also tested to see if the drug can be administered in different way and if it can be used in different stags of the disease and still work successfully.

3. RESULT

The results from these tests of potential drugs has not been 100% successful, there is a cure in terms of symptomatic cures/ treatments but there has been no break through in the terms of a disease modifying drug.

4. CONCLUSION

Due to repeated failures, it may suggest that we may not understand the disease completely and that we may have to go back to square one and try to understand completely.

REFERENCES

Mangialasche F., Soloman A., Winblad W., Mecocci P., Kivipelto M. et al. (2010) Alzheimer's Disease: clinical trials and drug development. *The Lancet Neurol 2010 Vol.9: 702-716*

FINDING WEAKLY INTERACTING MASSIVE PARTICLES

COOPER D P.

ABSTRACT

For many years, physicists have been trying to find explanations; Weakly Interacting Massive Particles (WIMPs) is one of these explanations. Until recent years these particles have just been hypothetical but the Cryogenic Dark Matter Search (CDMS) has found evidence that these particles are real.

1. HYPOTHESIS

The hypothesis is that these particles were of mass ranging from 1-100 GeV and only interacted with gravity and weak force. They also do not interact through electromagnetism which means they cannot be seen directly. The fact that this particle has these properties closely relates it to a neutrino; the only difference is that WIMPs are much more massive and are therefore slower. It is said that these particles are around about 50x larger than a proton.

2. TEST

The CSDM placed 11 Silicon- based detectors and 19 Germanium – based detectors in the Soudan mine, Minnesota. It was predicted that these detectors could record the extremely rare occurrence of a WIMP colliding with an atomic nucleus. The reason as to why Germanium was used is because it has a nucleus that is larger than a WIMP and so is more likely to have the particle collide with it but also protons, neutrons and all other smaller particles can collide with it too. The reason as to why the silicon detector was use is because it is large enough to have protons, neutrons and all other smaller particles collide with it but not large enough to successfully detect a WIMP. So by using the equation, Germanium event – Silicon event, you can work out the total WIMP events. The Germanium crystal in the detectors are chilled at a constant 50mK and was surrounded by tungsten and aluminium which are held at a critical temperature so that they in a superconducting state. The reason why the apparatus is set up like this is because when a WIMP passes through the Germanium crystal, there is vibration which will then cause a change in temperature. Due to the metal being superconductive there will be a change in resistance in the metal which is then recorded; different particles cause different amounts of vibrations and so different readings.

3. RESULT

In a recent gather of results CDMS stated there were 99.8% sure they had witnessed WIMPs of mass 8 GeV, showing a mass of proton to mass of WIMP ratios of 1-8 (E., 2012). However, this percentage of certainty is not high enough in terms of particle physics; they must be 99.9999…% in order for the results to be trusted.

4. CONCLUSION

In conclusion, if more of these results come back and are consistent each time of the mass and speed being the same, then the mystery of dark matter will have been solved.

REFERENCES

E., B. (2012, April 16). *Sci-guy*. Retrieved June 27, 2013, from Chron: http://blog.chron.com/sciguy/2013/04/scientists-quietly-announce-a-potentially-huge-discovery-in-physics/

3-D PRINTING FOOD IN OUTERSPACE

CROSSLEY.L

ABSTRACT

NASA has very recently provided Systems & Materials Research Corporation with a $125,000 grant to spend six months building a prototype of a 3-D food printer- this printer isn't your everyday printer, it'll be able to print out a tasty pizza. The main purpose is to eliminate any preserved foods in which astronauts take with them on their long haul space flight and replace them with the foods which they are used to back home. It aims to reduce the need to send space shuttles in to space delivering fresh foods for the astronauts, which will therefore save both money and time. It will also increase the health of astronauts as all the essential vitamins and nutrients are just a click away.

1. HYPOTHESIS

An advanced Food technology project scientist at NASA, worries for the future of space missions. He states that if food on long-haul space flights does not meet certain requirements, then the astronauts could become at risk from health problems and therefore jeopardising the whole space mission. Grace Douglas says "If they don't want to eat it, they won't eat enough"

2. TEST

Testing this food printing process takes several stages, the first stage was carried out by a team from NASA who demonstrated that the 3D printers can print in microgravity. The actual launch into space is set for 2014, "Next year, we will demonstrate that they can print on the International Space Station." - Grace Douglas

3. RESULT

Last November a prototype of Contractor's design was tested. It successfully printed a 3-D bar of chocolate. This therefore demonstrates that the already existent 3-D printers are capable of printing food already, however the next issue raised is whether the printer is able to print in zero gravity.

To prevent wasting both time and money NASA decided to test the product on earth using microgravity. They did this during four airplane flights that achieved brief periods of microgravity via parabolic manoeuvres. This test was completed on the 19th June 2013, and the printer was again successful.

4. CONCLUSION

To conclude, the development of this 3-D food printer is still underway, and is yet to be launched into space. Looking at the results from tests already completed this project has already been successful and should hopefully continue this way. By the end of this procedure, astronauts should be launched into space with a 3-D printer rather than preserved foods.

REFERENCES

(3D Printer Passes Zero-Gravity Test for Space Station Trip) (NASA asks: Could 3-D-printed food fuel a mission to Mars?) (Why NASA Just Spent $125,000 To Fund A 3-D Pizza Printer Prototype) (Home-baked idea? Nasa mulls 3D printers for food replication)

http://www.fastcoexist.com/1682194/why-nasa-just-spent-125000-to-fund-a-3-d-pizza-printer-prototype

Big Brain: An Ultra-High Resolution 3-D Roadmap of the Human Brain

DEXTER M.

ABSTRACT

The brain has been scanned before, however never in the microscopic detail that the Big Brain project offers. They sliced a 63 year old woman's brain into 7404 histological sections. The sections were then processed into a supercomputer to create a 3D model of the brain in high detail. The model of the brain named 'Big Brain' is freely available for neurologists and scientists to access online. The model allows for neuroanatomical insight in the brain which could allow in the future for specific parts to be inspected and emotions, languages and cognitive processes. This model is a huge improvement on the past models of the brain made by Brodmann and von Economo.

1. HYPOTHESIS

The aim of this study was to see if they could create the highest detailed model of the brain so far. The model was created to be in higher detailed than previous models suggested by Brodmann and von Economo.

2. TEST

During the test and the actual construction of the 3D model of the brain, the researchers took a 63 year old, epileptic woman who donated her brain to science. Using a machine called a microtome they were able to control the thickness of the sections of her brain. (Amunts, et al., 2013) They took 7404 sections of her brain in 20micrometres in thickness. The sections were stained so the cell structure could be identified before scanning it through a super computer. This process took in total 1000 hours to process. The computer was even able to construct the model together even if the slices were cut at an angle. In total the computer stored 1 trillion bytes of data on the 3D model of the brain.

3. RESULT

The result is the highest detailed brain ever created which will allow scientists to study in depth and the processes of the brain; this could also help in such things as helping those who suffer with Parkinson's disease as it will allow neurosurgeons placing electrodes for deep brain stimulation which can help stop or reduce the shaking. The model was 50 times higher in resolution than the highest resolution of the most powerful brain scan to date.

4. CONCLUSION

Overall the experiment was successful and its applications in the future will provide helpful research into how the brain works and possible cures for diseases that effects or is regulated in the brain.

REFERENCES

Amunts, K., Lepage, C., Borgeat, L., Mohlberg, H., Dickschied, T., Rousseau, M.-E., et al. (2013). *BigBrain: An Ultrahigh-Resolution 3D Human Brain Model.* Amunts, K.

TIN ANODE FOR SODIUM-ION BATTERIES USING NATURAL WOOD FIBER AS A MECHANICAL BUFFER AND ELECTROLYTE RESERVOIR

DEXTER M.

ABSTRACT

Sodium ion batteries are cheap, due to their abundance. Tin also has a high capacity for energy, so using these two elements together can make a cheap, energy efficient battery. However, a tin and sodium alloy can create expansion, and unstable solid electrolyte interphase, which effects conductivity. To overcome this, researchers must use a soft porous material to release the stress put upon the battery. Wood was found to be the perfect material as the structure of the wood allows ions to transport in and out of the battery. This has been confirmed both in experiments and computer simulation allowing the battery to recharge up to 400 times.

1. HYPOTHESIS

If using wood in a sodium ion and tin battery reduces the swelling in the batteries and makes the solid electrolyte interphase stable thus allowing it to conduct, due to its structure and porous materials. Meaning that is could withstand more charges than a regular sodium ion and tin battery. The wood would act as a buffer to reduce the strain put onto the battery during the sodiation/desodiation process.

2. TEST

They tested this both by computer simulation and actual experimentation, both times proving successful. To do this they worked on a microscopic scale. They coated cellulose microtubules around 25micrometres thick with a thin layer of tin approximately 50nm thick. This allows the sodium ions to freely move in and out of the wood due to its porous structure without damaging the anode. (Zhu, et al., 2013)

3. RESULT

It caused the cellulose to shrink and shrivel rather than swell and like regular sodium ion and tin batteries. Due to the strong structure of the cellulose in wood, it meant that is would not burst, but rather change shape and shrivel to deal with the stress and pressure put upon the conductive fibre during the sodiation and desodiation process. This is where the sodium ions move freely in and out of the conductive fibre, assisting in carrying the charge. The result has been to create a Nano-battery that is environmentally benign, long lasting due to the many charge it is able to hold and efficient.

4. CONCLUSION

In conclusion the researchers have found that wood helps to overcome the problems that a sodium ion and tin battery produce and therefore create a battery that can hold many charges and can continue to be reused.

REFERENCES

Zhu, H., Jia, Z., Chen, Y., Weadock, N., Wan, J., Vaaland, O., et al. (2013). *Tin Anode for Sodium-Ion Batteries Using Natural Wood Fiber as Mechanical Buffer and Electrolyte Reservoir.* Hongli Zhu.

EXPLOSIONS ON THE MOON

DOE B L

ABSTRACT

For years, man has looked up and observed the moon. Technology developed over time, as did the capability to see the moon closer through telescopes. Many astronomers who observed the moon observed small flashes on the lunar surface. For about a thousand years astronomers have noted several sightings of transient lunar phenomenon, which is a short lived change in on the surface of the moon (Wikipedia, 2013). NASA decided that they would train and record the lunar activities and attempt to observe any of these "transient lunar phenomena. So what were the flashes recorded? Meteorites hitting the lunar surface created explosions which would appear to be one to three second flashes when viewed from earth.

1. HYPOTHESIS

For about a thousand years, transient lunar phenomenon has been witnessed by astronomers and lunar observers have seen small, short-lived flashes on the moon. The lunar observers were not believed (Wikipedia, 2013). NASA took on the task of seeing if these phenomena are true so they set about watching the moon for changes on the surface of the moon. What they were monitoring on the lunar surface were any signs of meteorite explosions.

2. TEST

For 8 years, NASA has used 14-inch telescopes to record the lunar surface to look for signs of explosions. The telescopes are put outside in a waterproof dome so that they can get 360° view of the moon. The telescopes would record the data then analysts would watch and record characteristics of the meteorite. They would calculate the mass from the size of the crater left behind and the velocity in which the meteorite hit the moon. (Phillips, 2013)

3. RESULT

The meteorites hitting the earth gave of a small flash of light. However there is no oxygen on the moon, so why is there a flash? The flash is from the thermal glow of the molten rock. The rock melts due to the kinetic energy to meteorite has when it hits the lunar surface which is then transferred to thermal energy and melts the rock giving the glow. The Earth and the moon are constantly bombarded by meteor showers. On 17th March 2013, the largest explosion was witnessed with the brightness of a 4th magnitude star and the energy of 5 tons of TNT. The meteor hit the moon at 56,00mph. (Phillips, 2013).

4. CONCLUSION

They found that the flashes were meteorites hitting the moon. The earth and the moon are also constantly bombarded by meteor showers.

REFERENCES

Phillips, T. (2013, May 17). *Explosions on the Moon*. Retrieved from NASA: http://science.nasa.gov/science-news/science-at-nasa/2013/16may_lunarimpact/

Wikipedia. (2013, May 20). *Transient Lunar Phenomenon*. Retrieved from Wikipedia: http://en.wikipedia.org/wiki/Transient_lunar_phenomenon

TO SEE WHETHER MIRROR NEURONS CAN EXPLAIN PHANTOM LIMBS AND A METHOD OF HELPING THE SYPTOMS.

ESSOR.A

ABSTRACT

Researched information into mirror neurons: which are involved in empathy, understanding speech and many more things, and how they relate to helping patients who have experienced phantom limb pain, by coming up with a method to relieve symptoms of phantom limb pain and the effectiveness of this therapy.

1. HYPOTHESIS

The hypothesis was to see whether phantom limb pain can be relieved through patients using mirror visual feedback therapy. Mirror neurons were first discovered by Rizzolatti et al and they found that these neurons were involved when the animal observed and also, when the animal did an action (M.Glenberg, 2011).Ramachandran then looked as these mirror neurons and found that they could be a way of dealing with phantom limb pain. He wanted to see if this method could reduce symptoms of phantom limb pain such, as pressure, pins and needles, tingling or stabbing pain on the missing limb. He used a box with a mirror, and the patients phantom limb was behind the mirror so that when they looked at the mirror they saw the reflection of their actually to be seen as the missing limb. (BBC World Service, 2011)

2. TEST

He tested the mirror visual feedback therapy on a number of patients who had an amputation and also were experiencing phantom limb pain. They tried the Mirror Visual Feedback Therapy on a total of 10 patients. E.g. one patient called R.T. had one of his arms amputated 7 months before meeting up with Ramachandran. When he looked into the mirror he felt that he could actually create voluntary movement in his arm again. (W.Hirstein, 1998)

3. RESULT

Ramachandran found that in 6 of the 10 patients that the phantom limb moved with the mirror image however, this didn't happen to the other four patients. And, by repeating the process or the M.V.F.T, it led to a disappearance of the phantom limb altogether.

4. CONCLUSION

I found that through understanding the properties of mirror neurons that they have helped find a method of reducing phantom limb pain by using a mirror visual feedback therapy and by watching another person having a massage or flexing can also relieve the pain which is experienced with a phantom limb. And this method seems to be the most effective way of reducing phantom limb pain.

REFERENCES

Arthur M.Glenberg. (2011). Introduction to the Mirror Neuron Forum. *Perspectives on Psychological Science 2011 6:363*

V.S. Ramachandran and W. Hirstein. *The perception of phantom limbs*. The D.O. Hebb Lecture. *Oxford University Press (1998)*

V.S. Ramachandran and D. Rogers-Ramachandran. *Phantom limbs induced with mirrors. (1996)*

BBC World Service. (2011). W*hat phantom limbs and mirrors teach us about the brain*. http://www.bbc.co.uk/news/magazine-15938103

THE PERFORMANCE OF A REWALK
GOODBAND H

ABSTRACT

The research I under took was, to see how efficient and safe the ReWalk machine is and how it helps people with thoracic spinal cord injury walk again. The robotic machine has been recently released to the public at a cost of $85,000. Before its public release testing had to be done to ensure its safety when using it and also to know the limits of the robotic machine as not all paraplegics are able to use the robotic machine. 12 individuals have taken part in testing the ReWalk.

1. HYPOTHESIS

The test I researched was to study the performance and safety of the ReWalk exoskeleton (November 2012 - Volume 91 - Issue 11 - p 911–921). 12 individuals have been put through training with the ReWalk exoskeleton to see whether any implications have occurred so that improvements to the ReWalk exoskeleton could be made. Different types of paraplegics have also been used to test the ReWalk to see whether any other if different individuals with other spinal cord injuries are able to operate the robotic machine.

2. TESTING

After training with the ReWalk exoskeleton the individuals were asked to walk as far as they could and state any impactions they came across when operating the ReWalk exoskeleton.

3. RESULTS

After training, all subjects were able to independently transfer and walk, without human assistance while using the ReWalk, for at least 50 to 100 m continuously, for a period of at least 5 to 10 mins continuously and with velocities ranging from 0.03 to 0.45 m/sec. Excluding two subjects with considerably reduced walking abilities, average distances and velocities improved significantly. Some subjects reported improvements in pain, bowel and bladder function, and spasticity during the trial. All subjects had strong positive comments regarding the emotional/psychosocial benefits of the use of ReWalk. (November 2012 - Volume 91 - Issue 11 - p 911–921)

4. CONCLUSION

ReWalk holds considerable potential as a safe ambulatory powered orthosis for motor-complete thoracic-level spinal cord injury patients. Most subjects achieved a level of walking proficiency close to that needed for limited community ambulation. A high degree of performance variability was observed across individuals. Some of this variability was explained by level of injury, but other factors have not been completely identified. Further development and application of this rehabilitation tool to other diagnoses are expected in the future. (November 2012 - Volume 91 - Issue 11 - p 911–921)

REFERENCES

November 2012 - Volume 91 - Issue 11 - p 911–921
http://journals.lww.com/ajpmr/Abstract/2012/11000/The_ReWalk_Powered_Exoskeleton_to_Restore.1.aspx

DEEP BRAIN STIMULATION AND PARKINSON'S SWALLOWING FUNCTION

GULLEY C

ABSTRACT

My presentation is looking at the research into DBS and how it can help Parkinson's patients swallow. The research was carried out by Kulneff L et al to reduce the chances of aspirational pneumonia which is the most common cause of death in PD.

1. HYPOTHESIS

Parkinson's is a degenerative brain disease which affects the patients with tremors, rigidity, postural instability, depression and dementia and other symptoms. The most common cause of death in PD patients is aspiration pneumonia. DBS is already a proven treatment for PD on motor controlled symptoms such as tremors the research was carried out to see if DBS will be an effective treatment for the swallowing dysfunction.

2. TEST

The test was carried out on 11 patients aged 41-72. The patients had electrodes implanted in the subthalamic nucleus and then were stimulated. The Patients were the monitored using self-estimation and fiberoptic endoscopy. These were carried out after 6 months and then 12 months.

3. RESULT

The results of the test showed that the DBS had no negative effects on the swallowing function according to the fiberoptic endoscopy but the self-estimation showed that there was a dramatic improvement in the swallowing.

4. CONCLUSION

In conclusion DBS had no negative effects on the swallowing function in PD patients.

REFERENCES

Kulneff L, Sundstedt S, Olofsson K, van Doorn J, Linder J, Nordh E, Blomstedt P.

HORSES (EQUUS CABALLUS) RECOGNISE KNOWN HANDLERS.

HALL S.

ABSTRACT

It was shown domestic horses recognise and distinguish between members of different species like humans. This was tested in 2 parts. The first showed horses a known handler and a stranger. Simultaneously recordings of both people calling the horse's name were played one after another. The horse correctly matched the voice of their handler to that individual. In the second test, they were shown 2 known handlers; following the same process but the handlers' voices were played`. This showed, horses could match each handler to the correct voice. The horses performed better with the handler on the right side indicating the left hemisphere of the brain is dominant in matching visual and aural information.

1. HYPOTHESIS

Professors at the University of Sussex (Proops & McComb, 2012) suggested horses may be capable of cross-modal recognition (using a combination of auditory and visual information to form mental images of individuals, meaning they could match the voice of a handler to the correct person) of individuals from species genetically different from their own. It was proposed there may be bias in the involvement of each hemisphere of the brain.

2. TEST

The theory was tested in 2 parts. The first, asked if horses could distinguish between familiar and unfamiliar humans. A known and an unfamiliar voice played from a hidden loud speaker, calling each horse's names; the known handler and stranger stood each side, 6 metres from the speaker, facing the horse. This was done with 32 horses. The second experiment tested horses' ability to distinguish between 2 known humans. 39 different horses were tested, ensuring no corruption of results due to associations perhaps made in the first test. The horses were shown 2 known handlers, each voice, calling the horse's name was played. The responses of horses in both tests were videoed and analysed frame by frame.

3. RESULT

The first test showed horses looked for longer time periods at a known handler when their voice played, whereas the horse would look side to side not distinctly in any direction when the strangers' voice played, demonstrating no connection between horse and stranger. The second test showed horses could match each handler to the correct voice, due to the animal always looking toward the human whose voice was playing for a longer time period than at the person whose voice was not playing. Horses performed better when the person whose voice was playing stood to the right. Mares were better at performing the matching task than geldings.

4. CONCLUSION

The research confirms domestic horses form visual and auditory associations with known handlers and they use this ability to distinguish between 2 handlers. The sidedness of the horses means they could be trained from the right not the traditional left side. The use of mares may increase.

REFERENCES

Proops, L., & McComb, K. (2012). cross modal individual recognition in domestic horses (equus caballus) extends to familiar humans. *Proc. R. Soc. B 22 August 2012 vol. 279 no. 1741 3131-3138.*

THE TRAINING OF INSECTS WITHIN THE ORDER "HYMENOPTERA" TO DETECT SUBSTANCES

HARTOP G.

ABSTRACT

Currently, Canines are recognised as detection animals world-wide; however, with their low rate of successful detections at roughly 30% (Appel, 2010), high cost of training and maintenance, could other methods be the way forward for substance detection? One method for substance detection involves using a species of parasitic wasp, *Microplitis croceipes*. This type of wasp can detect a chemical in caterpillar faeces and track the caterpillar, before injecting its eggs which grow and feed on the caterpillar. The high sense of smell from the wasp could be used to detect chemical scents within the air with greater succession than that of canines.

1. HYPOTHESIS

If the olfactory system in *Microplitis croceipes* is harnessed, trace vapour detection of substances could be made more reliable, easier and cost-effective. Insects could be trained within a few minutes to respond to a specific odour. Through associative learning, wasps can link certain odours to a host or food, much like Pavlov's dogs associated a ringing bell with food, bees, and wasps can link odours with a food. If wasps are trained to give a behavioural response upon odour detection, then the whole process of screening at secure areas and forensic screening of a scene can be made quicker, easier and less costly.

2. TEST

A model species was selected for this test, *Microplitis croceipes*, to determine the threshold of insect response for four compounds: 3-octanone (a compound which is found in numerous fungal pathogens), myrcene (a volatile compound released by cotton plants when eaten by cotton bollworms), putriscene and cadaverine (two products released by the breakdown of dead animal protein by microorganisms). Eighteen test wasps were trained to each of the individual compounds at one dose. Doses were then decreased, until the responses were negligible. The response to the odour was defined by a searching behaviour called antennating. Response was measured by the length of time the wasp antennated while exposed to the odour. (Rains, Tomberlin, D'Alessandro, & Lewis, 2004)

3. RESULT

The mean wasp response fell below 10s at approximately 3.1×10^{-7}, 2.9×10^{-7}, 3.9×10^{-6}, and 4.5×10^{-7} mol L^{-1} of compound for 3-octanone, myrcene, cadaverine, and putriscene, respectively.

4. CONCLUSION

For comparison, the detection limits of an electronic nose, the Cyranose 320, was determined for two of the four compounds. The response limits of the wasp for the compounds 3-octanone and myrcene were 74 and 94 times better than the electronic nose, respectively. The response limit of the wasps to putriscene, 3-octanone, and myrcene was approximately 10 times better than to cadaverine.

REFERENCES

Appel, A. (2010, October 28). Drug-Sniffing Wasps May Sting Crooks. *National Geographic News*, pp. 1-2.

Rains, G., Tomberlin, J., D'Alessandro, M., & Lewis, W. (2004). Limits Of Volatile Chemical Detection Of A Parasitoid Wasp, Microplitis crocepes, And An Electronic Nose: A Comparitive Study. *Transactions of the ASABE*, 2145-2152.

EARLY CDT- LUNG CANCER DETECTION

HICKENBOTHAM. D

ABSTRACT

Lung cancer is currently the second most diagnosed cancer with 113 people being diagnosed with the cancer each day. The EarlyCDT was invented and developed by a Professor of Surgery at Nottingham University. It was designed to detect cancers earlier than conventional screening and to improve the effectiveness of treatment. It specifically tests for cancer of the colon, lung, prostate and breast. In my presentation the aspects that I will explore are why the test is needed, how it works, the results from the test and what this means as well as the facts that show the effectiveness of the test.

1. HYPOTHESIS

The hypothesis for this test was to see if having a technique available for detecting cancer within the earliest of stages can make a difference in the amount of people who can receive treatment and overcome ling cancer. It was also to see how successful this technique was. Following this you would expect to find that earlier detection could decrease the amount of deaths from Lung cancer as treatment can be received earlier in order to destroy the cancerous proteins in the patients' blood.

2. TEST

A test was carried out on 69 patients who were considered at high risk of lung cancer to see what changes the cancer makes to different parts of the body including the blood.

3. RESULT

During the series of tests that they did they carried out blood tests on each of the patients and found that 57% of the patients had cancer associated auto-antibodies in their blood sample, 50% of these were before they had even been diagnosed with lung cancer. And, 15% of these didn't have lung cancer or any evidence of the disease.

4. CONCLUSION

The Early CDT-Lung test is 91% accurate and has the highest specificity levels out of all of the other cancer screening methods. As the test can detect a single protein in the blood it allows for cancer to be detected 5 years before conventional screening allowing for treatment to be given quicker. Finding the cancer at its earliest stage also means that the survival rate is substantially higher. The test will also have a huge impact on the future of survival rates of

	SENSITIVITY	SPECIFICITY	PREVALENCE YEAR RISK (1ST YR)	ACCURACY	PPV RATIO
EARLYCDT-LUNG	41%	93%	2.0%	91%	1 in 5
CT	70%	50%	2.0%	50%	1 in 36
MAMMOGRAPHY Women <50 years	40%	92%	0.8%	92%	1 in 26
PSA	33%	86%	2.4%	85%	1 in 18

lung cancer as early detection of the cancer means that less aggressive forms of treatment will be required. As the test has no side effects and minimal risk it means that more people will be more likely to have this test rather than the conventional screenings. Ultimately the impact of this test is to save lives that could possibly lost due to late detection of lung cancer.

REFERENCES

http://www.channel4.com/news/breakthrough-cancer-bloodtest-to-be-made-available-in-uk

http://www.huffingtonpost.com/2011/01/03/a-new-blood-test-to-detec_n_803462.html

http://scholar.google.co.uk/scholar?q=early+cdt+test&btnG=&hl=en&as_dt=0%2C5&as_ylo=2009

http://www.earlycdt-lung.co.uk/what-is-earlycdt/

http://www.earlycdt-lung.co.uk/learn-more/brochures_for_medical_professionals/ -Frequently asked question (FAQ's) Document for Doctors EarlyCDT-Lung

http://www.nottingham.ac.uk/impactcampaign/campaignpriorities/healtndwell-being/cancerearlydetection/cancerearlydetection.aspx

THE AFFECT OF IRON TABLETS ON MATERNAL ANAEMIA AND BABIES BIRTHWEIGHT

IRSHAD H

ABSTRACT

Haider from the Harvard School of Public Health led a team of researchers in Britain and the USA to conduct clinical trials and observational studies involving nearly 2 million pregnant women. They wanted to investigate from this study the link between prenatal iron use, maternal anaemia and low birth weight of babies. The researchers used a systematic review; considering the significant points of current knowledge focusing on the research question making all high quality evidence relevant to that question. They also used meta-analysis so the results could be compared easily from the two different studies hoping to find a relationship in their data. Iron deficiency is the most widespread nutritional deficiency in the world (Stoltzfus R, 1998) and the most common cause of anaemia during pregnancy (WHO) especially in low and middle income countries. During pregnancy a woman's blood volume increases by 50% so more iron is required for more haemoglobin so more oxygen can be carried to the foetus and mothers tissues.

1. HYPOTHESIS

Researchers aimed to investigate using a systematic review and meta-analysis whether there is a link between use of iron pills during pregnancy, maternal anaemia and birth weight of babies. They looked at high and low income countries to make results more valid.

2. TEST

Using databases like PubMed and Embase which published studies up to May 2012 Haider's team (Batool A Haider, 2013) conducted their research. Clinical trials of 48 randomised control trials consisting of 17,793 women were conducted. They investigated the use of the daily iron pill or a placebo then examined maternal haemoglobin levels to see if lower than 110g/l they would be considered anaemic and measured babies birth weight. Observational studies consisting of 1,851,682 pregnant women were conducted of which there were 44 cohort studies. This research prospectively followed the relationship between baseline anaemia and birth weight.

3. RESULT

The clinical trial results show that from 36 trials iron increased the mothers' haemoglobin concentration by an average difference of 4.59g/l. From 19 trials iron reduced the risk of maternal anaemia by 50%. Researchers estimated for every 10mg increase of iron per day to 66mg/day the risk of maternal anaemia by 12% and the birth weight increased by 15.1g. Mothers taking iron pills had a 19% reduction in risk of having a low birth weight baby and babies were 41.2 g heavier than mothers given placebo. Results of the observational studies found iron deficiency in the first 6 months of pregnancy created a 29% higher risk of having a low birth weight baby.

4. CONCLUSION

The researchers concluded that daily iron supplements during pregnancy decrease the risk of maternal anaemia and having a low birth weight baby by improving haemoglobin concentrations and preventing growth retardation of the foetus.

REFERENCES

Batool A Haider, I. O. (2013). Anaemia, prenatal iron use, and risk of adverse pregnancy outcomes: systematic review and meta-analysis. *British Medical Journal*, 346.

Stoltzfus R, D. M. (1998). *Guidelines for the use of iron supplements to prevent and treat iron deficiency anaemia.*

WHO. *Micronutrient deficiencies: iron deficiency anaemia.* Retrieved from World Health Organization.: http://www.who.int/nutrition/topics/ida/en/

SCALE: A CHEMICAL REAGENT THAT TURNS MICE BRAINS TRANSPARENT

IRSHAD H

ABSTRACT

Researchers at RIKEN, have developed a revolutionary new aqueous reagent which literally turns biological tissue transparent. Experiments using fluorescence microscopy on samples of mouse brain treated with this reagent have produced 3D images if neurons and blood vessels. Highly effective and cheap to produce, the reagent offers an ideal means for analysing the complex organs and systems that sustain living organisms. Scale is an improvement of other clearing reagents as it does not alter the overall shape of the brain, avoids light scattering and penetrates through a few millimetres rather than just a few micrometres. The chemical reagent also decreases the intensity of signals emitted by genetically coded proteins in the tissues which are used as markers to label specific cell types. The researchers have already studied neurons and blood vessels in the mouse brain in very high resolution.

1. HYPOTHESIS

Khayet and Matsuura (Matsuura, 2001) found polyvinylidene fluoride membranes became transparent when they soaked up 4M of urea. So RIKEN researchers aimed to investigate whether they could produce a chemical clearing reagent which does not have the limitations of past reagents to view neurons in a mammalian brain.

2. TEST

The researchers (Hiroshi Hama1 H. K.-S., 2011)treated mouse brain sections 60μm thick fixed with 4% paraformaldehyde with solutions containing 1-8M of urea. After 48 hours they treated sections with 4-8M urea turning it transparent along with some expansion. Then they combined the solutions with other ingredients. The most effective was colourless 'Scale A2'composed of 4M urea, 10% wt/vol glycerol and 0.1% wt/vol Triton X-100. A quantitative measure of the transparency was taken using transmission of light using a spectrophotometer. Then incubated brain slices in Scale A2 solution and less than 2 weeks later the slice was transparent compared against a patterned background. Scale U2 was developed which reduces expansion and fragility of samples. Brain samples were washed with 20 volumes of PBS. After 2 sequential washes for 15 minutes each the samples returned to their original structures.

3. RESULT

The results show that scale A2 transmits the most light of 350-920nm so is the best biological clearing agent. The expansion can be prevented by using scale U2 which has contains more glycerol to prevent excess hydration. Neurons and blood vessels in the brain can be seen using this reagent and magnifying.

4. CONCLUSION

The researchers concluded that they had found the most effective biological clearing reagent yet which can help future research using human and other organs biopsies.

REFERENCES

Hiroshi Hama, H. K.-S. (2011). Scale: a chemical approach for fluorescence imagingand reconstruction of transparent mouse brain. *Nature Neuroscience*, 1481-1489.

Matsuura, M. K. (2001, October 21). Preparation and Characterization of Polyvinylidene Fluoride Membranes for Membrane Distillation. pp. 5710-5718.

HIGGS BOSON

KATSAROS A.

ABSTRACT

The Higgs Boson was initially theorised by Peter Higgs in 1964. His theory was that there is an elementary particle which gives causes some fundamental particles to have mass when they should in fact be massless. The Higgs Boson is a very important discovery as it appears to confirm the existence of the Higgs Field.

1. HYPOTHESIS

The Higgs Boson is thought to confirm the existence of the Higgs Field, which fits in with the Standard Model and other theories within particle physics.

2. TEST

The test for this was carried out at the Large Hadron Collider at CERN in Geneva, Switzerland. The Large Hadron Collider fired two nucleuses from lead atoms at speeds near the speed of light and made them collide. This had to be done all day every day to collect enough data to confirm the existence of the Higgs Boson, as only 1 collision in 1 billion creates a Higgs Boson. Because of how rare the creation of a Higgs Boson is, and many other possible collision events can have similar decay signatures, the data of hundreds of trillions of collisions needs to be analysed and must "show the same picture" before a conclusion about the existence of the Higgs boson can be reached (Higgs Boson, 2013).

3. RESULT

The test confirmed the existence of a Boson which fit in with the theory made by Peter Higgs, and after almost a year of analysis this was confirmed to be the Higgs Boson. The analysis took a long time because the Boson found had a mass of 125 GeV, but the standard model suggests that it's mass should be infinite. They realised that this means there is more physics to be discovered. The discovery of the Higgs Boson means that the existence of the Higgs Field, a field which covers the entire universe and causes everything to have mass, is very feasible.

4. CONCLUSION

The Higgs Boson is an elementary particle which causes all things to have mass even though they physically shouldn't. Its existence confirms the existence of the Higgs Field, which explains why all things in the universe has mass.

REFERFENCES

Higgs Boson. (2013, March 28). Retrieved June 27, 2013, from Wikipedia: https://en.wikipedia.org/wiki/Higgs_boson#Search_prior_to_4_July_2012

WATER VS. SPORTS DRINKS

LYNCH E.

ABSTRACT

Being hydrated is very important in your everyday life, especially during physical activity. There is a large amount of evidence showing that exercise-induced dehydration has a negative impact on exercise performance and restoration of fluid balance must be achieved after exercise. Therefore, during physical activity, a discussion will undertake to whether isotonic sports drinks such as Lucozade Sport are as beneficial as it states, compared to those of ordinary tap water.

1. HYPOTHESIS

It has been suggested from previous evidence that drinking during exercise improves performance, provided that the exercise is of a sufficient duration for the drink to be emptied from the stomach and be absorbed in the intestine. Generally, drinking plain water is better than drinking nothing, but drinking a properly formulated carbohydrate electrolyte sports drink will allow for an even better exercise performance. Therefore the new hypothesis is to see whether drinking an isotonic sports drink is more beneficial to you within hydration, compared to that of ordinary tap water. (Susan M. Shirreffs, 2003)

2. TEST

A group of individuals were given several different drinks with different compositions which include carbohydrates, sodium and potassium. Their activity carried out for no longer than about 30–40 minutes each. To see whether those with compositons of those in a 'typical' sports drink, performed better with those who just had water. The test was highly controlled, doctors and coaches were aware of the electrolytes lost in sweat for different people.

3. RESULT

In the experiment, they found that sodium is the most beneficial electrolyte within hydration, thus the reason to why it is contained in all isotonic sports drinks

4. CONCLUSION

Sodium is key within hydration, and is ultimately the reason to why isotonic sports drinks contain the most sodium, to help replenish rehydration.

REFERENCES

United States Patent – Elseviers et al (Oct. 2, 2001)

The Optimal Sports Drink – Susan M. Sherreffs

Sports Nutrition Vlog (Water) – Mark Lovell

www.bmj.com/content/345/bmj.e4737?hwoasp=authn%3A1360059874%3A128%3A525547310%3A0%3A0%3ArzNwNVjirH3V0UYy0MlhsQ%3D%3D

VITAMIN B PREVENTS ALZHEIMER'S DISEASE

MCLEOD K.

ABSTRACT

Researchers at the Universities of Oxford, Warwick and Oslo investigated the prevention of Alzheimer's disease, through vitamin B supplements among patients known to be suffering from mild cognitive impairment. Alzheimer's disease is the weakening of mental activity, commonly linked with memory and language processes. Alzheimer's disease accelerates the natural process of the brain shrinking with age. It is a degenerative disease commonly occurring in elderly and middle aged patients. The disease occurs through degeneration of the brain, where plaques of protein start to develop, resulting in the death of the brain's nerve cells.

1. HYPOTHESIS

The researchers aimed to discover if there was an effect of vitamin B on the shrinking of the brain, in areas associated with Alzheimer's disease; in the cerebellum. (Gwenaëlle Douauda, 2013) The trial hoped to see that vitamin B supplements prevented the decrease in size of grey matter (nerve cells which do not have protection from myelinated cells). Research on the effects of homocysteine levels was also investigated alongside the trial.

2. TEST

156 elderly volunteers participated in a randomised controlled trial to see the effect of vitamin B on the reduction in size of grey matter. The volunteers who participated in the clinical trial fulfilled the criteria of suffering from mild cognitive impairment and being 70+ years old. The volunteers were randomly split into two equally sized groups and were selected to receive, or to not receive, the vitamin B treatment (folic acid 0.8 mg, vitamin B6 20 mg, vitamin B12 0.5 mg) whilst the other group received a placebo. The volunteers each had a magnetic resonance imaging of the brain at the start of the study. These images were then compared with a new MRI at the end of the study, after 24 months, to see if the vitamin B treatments had prevented the shrinking of grey matter in the brain. Homocysteine levels, in blood samples, were also measured at the start and end of the study.

3. RESULT

The size of grey matter at the start of the study was similar in both groups. Over the 24 months period, the area of grey matter had shrunk in both the vitamin B and placebo receiving groups. The volunteers who had received vitamin B supplements had a decrease in grey matter less than the placebo receiving volunteers. It is reported that there was a reduction in the loss of grey matter associated with the regions of the brain most affected by Alzheimer's disease, in both groups. The research suggests vitamin B treatment leads to a reduction in homocysteine levels as it is converted into acetylcholine (a chemical associated with memory functions), which slows down the rate at which grey matter is lost.

4. CONCLUSION

Vitamin B alone will not prevent Alzheimer's disease; however it will reduce the rate at which the accelerated shrinking of the brain occurs, when considered with other contributing factors.

REFERENCES

Gwenaëlle Douauda et al. (2013) Preventing Alzheimer's disease-related grey matter atrophy by B-vitamin treatment. *PNAS*.

A NEW PERSPECTIVE OF AN UNDISCOVERED BIDEPAL PRIMATE IN NORTH WEST AMERICA

MEACHEM B

ABSTRACT

Sightings of a bipedal primate have been seen throughout America and for hundreds of years, Native Americans have believed of its existence for hundreds of years with depictions of the animal in ancient cave art. Yet its existence it still met with trepidation and Sinicism by many scientists and researchers. The infamous Patterson Gimlin film of 1967 first brought to light the possibility of the existence of so called "bigfoot" which is believed to be the bipedal primate. The films along with photographed footprints have never been fully disproven, with all attempts to disprove the video and the individuals involved, Bob Gimlin and Roger Paterson remaining inconclusive.

1. HYPOTHESIS

New analysis (Meldrum, 2004) shows signs of primate morphology in photographs of footprints, which shows there is more evidence to support the Patterson Gimlin case, which is further reviewed (Meldrum, 2007), in which there was a video of around two minutes in length showing what appears to be a large bipedal primate. The footprints show signs of a mid-tarsal break which is a joint that is found in chimpanzee feet, this feature is a product of evolution that allows primates greater mid foot flexibility used for grasping which gives them a greater climbing ability. The footprint also has a proportionally larger heel that a human foot, this is another characteristic of primate feet, known as an elongated heel.

2. TEST

Jeff Meldrum undertook intensive 3D scans of many different footprints to compare and show repeating features of cast tracks. The 3D virtualisations of the footprint casts give a greater perspective of the mid tarsal break shown in the casts and allow precise measurements to be made.

3. RESULT

The result is that the casts show indefinite signs that the foot that created the print must have had a mid-tarsal break, which is a feature of primate morphology. The results also show that the cast has an elongated heel proportional the footprint. Both of these have been found by using measurement from the 3D virtualisation created by Jeff Meldrum and Idaho State University.

4. CONCLUSION

The results are that the footprints show signs of primate morphology, which supports Patterson Gimlins account at Bluff creek in 1967. However whether this could be created by hoaxers is still debated. Therefore the evidence of primate morphology in footprints cannot prove or disprove the existence of a bipedal primate in North America.

REFERENCES

Meldrum, J. (2004). Midfoot Felxibility, Fossi Footprints, and Sasquatch Steps: New Perspective on the Evolution of Bipedalism . *Journal of Scientific Exploration*, 0892-33.

Meldrum, J. (2007). ICHNOTAXONOMY OF GIANT TYARCKS IN NORTH AMERICA . *Spielmann and Lockley*, 225-31.

FRUIT JUICE INFUSED IN CHOCOLATE TO REDUCE FAT CONSUMPTION

NIDHI N.

ABSTRACT

Chemistry professor Stefan Bon et al from the University of Warwick found a way of making healthier chocolate by infusing chocolate with fruit juice from the research carried out by them. These juices included orange, apple & cranberry juice and also infusing chocolate with diet coke or vitamin C water to replace up to fifty per cent of the fat that is normally in chocolate. Bon's team used fruit juices and other ingredients to form a Pickering emulsion. Chocolate is an emulsion of cocoa butter combined with cocoa powder with either milk or water. Lecithin is an emulsifier which appears on the ingredient label in many chocolates as it develops the process. This method however used solid particles rather than an emulsifier.

1. HYPOTHESIS

Stefan Bon et al aimed to use various fruit juices to replace fat consumption within chocolate. They expected that this new, healthier theory would not change the taste nor the texture of the chocolate and instead it would just provide consumers with a healthier option. (Hughes, 2012)

2. TEST

Cocoa butter and milk fats are replaced with fruit juice in the form of emulsion droplets using a quiescent (state or period of inactivity) Pickering emulsion fabrication strategy. This is an emulsion that is stabilised by solid particles, which absorb onto the interface between two phases. The advantage of adding these solid particles is that when added to the mixture, they will bind to the surface of the interface and prevent the droplets from combining in mass or whole, thus causing the emulsion to be more stable. However if just oil and water are mixed and small oil droplets are formed and dispersed throughout the water, eventually the droplets will coalesce to decrease the amount of energy in the system. Fruit juice was infused in the form of micro-bubbles (Pappas, 2013) and these tiny bubbles are able to replace the fat without undoing the velvety "mouth-feel".

3. RESULT

Although some may argue that the taste would change, it was found that this approach maintains the things that make chocolate 'chocolaty' but with fruit juice instead of fat. Following on from this discovery, food industries may be encouraged to take further action into making lower-fat chocolate. This new, improved way of making chocolate is found to work for dark, milk and white chocolate. Furthermore, from the results it can be concluded that milk chocolate formulation with water containing a tiny bit of vitamin C had quite a similar taste to ordinary milk chocolate compared to the use of fruit juice or flat coke.

4. CONCLUSION

To conclude, the new possibility of replacing part of the fat in chocolate with water-based juice droplets allows for greater flexibility and tailoring of both the overall fat and sugar content.

REFERENCES

T S Skelhon et al, J. Mater. Chem., 2012

S. Pappas, Live Science, 2013

E. Hughes, Chemistry World, 2012

TREATMENTS OF JAUNDICE, PRESENT AND UPCOMING

PACK A

ABSTRACT

The topic researched was jaundice. The explanation of what jaundice is was discussed and what types of treatments are available to treat jaundice were researched. The areas explored were different types of treatment, this included phototherapy, an upcoming enzyme treatment and a biliblanket.

1. HYPOTHESIS

An hypothesis is a specific statement of prediction. It describes in concrete (rather than theoretical) terms what you expect will happen in your study. Not all studies have hypotheses. Sometimes a study is designed to be exploratory. There is no formal hypothesis, and perhaps the purpose of the study is to explore some area more thoroughly in order to develop some specific hypothesis or prediction that can be tested in future research. The research about jaundice does not have a hypothesis.

1. CONCLUSION

In conclusion, the main treatment for jaundice is phototherapy, it can be done in the hospital and can quickly cure the jaundice if it is not too severe. The biliblanket is used in more severe cases and can be used at home, which is good for parents as they would be able to bring their child home for treatment instead of staying in the hospital. The enzyme treatment being worked on will hopefully prevent jaundice occurring but also help treat it.

REFERENCES

http://www.parenting.com/article/newborn-jaundice-treatment

www.stjoes.ca/media/PatientED/P-T/Phototherapy-lw%20(3).pdf

http://www.emedicinehealth.com/phototherapy_for_jaundice_in_newborns-health/article_em.htm

http://www.mayoclinic.com/health/infant-jaundice/DS00107/DSECTION=treatments-and-drugs

http://www.hngn.com/articles/4885/20130610/new-discovery-researchers-lead-better-jaundice-treatments.htm

THE BIOCHEMISTRY BEHIND LOVE AND ITS EFFECTS.

PANCHOLI.P

ABSTRACT

The question of 'what is love' will have a different answer depending on whether a biologist, psychologist or theologian is answering. In terms of neuroscience, love is much stronger than a basic emotion or state of mind. It is anticipated than from the findings and a greater understanding of these chemicals, scientists could develop essentially a 'love drug' which could be beneficial for autistic people who find it difficult to build or respond to communication or relationships.

1. HYPOTHESIS

Neurobiologists have studied the brain activity to show roles for oxytocin, vasopressin, and dopamine and their receptors. They are proving that love is more than just a feeling and can be explained with a neurobiological or chemical equivalent neural circuit in the brain which essentially has certain chemicals which trigger the chemicals all working in a 'romantic system'. It is approximated that 80% of waking time is spent obsessing or repeatedly thinking about their 'partner'. (Tarlaci, 2012)

2. TEST

In order to support the hypothesis, numerous tests were carried out. An article (Smith, 2010) shows the effects on a mother sheep once given a short puff of oxytocin. The hypothesis was also tested on human volunteers. The same volumes of doses of oxytocin were administrated to the volunteers. Furthermore, 18 people who were passionately in love, were used in a fMRG study (Bartels and Zeki, 200; 2004) where they were shown a photograph of someone whom they love to investigate the brains response. The results were compared to a normal person's brain.

3. RESULT

The result from the first test was that the sheep was persuaded to foster a lamb who was not her own. This reiterates that oxytocin is associated with attachment and ability to develop trust for people and more sensitive to others. In human volunteers the oxytocin caused developments of enhanced trust sensations and more sensitive to the emotions of others. They also spent longer looking at peoples' faces all characteristic of a person who is in love. Vasopressin causes men to be more aggressive and possessive over partners. Dopamine causes attraction and the addictive behaviour so wanting to see your partner as often as possible. After looking at the fMRG images, the test subject's brain subcortical reward system showed great activity. Those regions which rewards, gave a bigger response to the experiment. A reward will then trigger a repetition of that activity, the same thing happens with addictive substances like cigarettes or cocaine or simple things like water and positive social interactions.

4. CONCLUSION

With a greater understanding of the brain activity and the importance of oxytocin, dopamine and vasopressin, it is hoped that scientists could one day develop a 'love drug' which could not just improve people's relationships and even those suffering with autism.

REFERENCES

Tarlaci (December 2012) The Brain In Love: Has Neuroscience stolen the secret of love?

Smith (2010) What is Love?

GLUCOSE SENSING CONTACT LENSES FOR DIABETICS

PANCHOLI.P

ABSTRACT

The concept of using lenses for diabetic monitoring has been around for years, as the glucose levels in tears can track blood glucose levels. There have been various developments in this area however it is anticipated that diabetics could in the future be able to monitor their blood sugar levels using bionic contact lenses, instead of the current process of pricking fingers. The concept of a powered electrochemical cell in the eye is cutting edge and idea of wearable electronics are to most unimaginable. It may well find other medical applications in future. (Stoye, 2013)

1. HYPOTHESIS

It was important to find a new more high tech approach to glucose monitoring. One approach to track blood glucose in tears is by using disposable and colourless contact lenses where you can colourimetrically see changes in the contact lens colour or other fluorescence-based properties. This gives an indication of tear and blood glucose levels. (Al R. B., 2005). A newer hypothesis is an electrical glucose sensor in a lens which could display on the-spot reading that could be easily read by the wearer. The problem with the hypothesis is because of the complexity of the sensor it would need a power source, and so far this has proven a major obstacle for scientists. However now scientists have developed a fuel cell which runs on tears and the chemicals within the tears could power lens-mounted glucose sensors. The ascorbate and oxygen being oxidised and reduced created an electrical current to fuel the cell.

2. TEST

Shleev at Malmo University, Sweden, (Al S. E., 2013) used human tear samples to show the cell could generate power from tears without altering their glucose content. A Three dimensional nanostructured electrode was made by modifying gold wires with 17nm gold nanoparticles which were then modified again to make an anode and cathode. This reacts with the oxygen in tears where each electrode either oxidises or reduces the oxygen.

3. RESULT

When human tears were used, the biodevice showed at 0.54 V a maximal power density of 3.1 uWcms and at 0.4V the current density output was over 0.55 for 6 hours of continuous operation showing that the ascorbate oxygen could be used a power source for the bio cell and used for continuous health monitoring for those suffering from diabetes.

4. CONCLUSION

An electrochemical cell in the eye and wearable electronics like this are innovative and ground breaking however as of now, the commercial ability of the theory is limited as the expense makes it less approachable for many. However the hypothesis may well find other medical applications in future and change the lives of diabetics for the better.

REFERENCES

S Shleev et al, Anal. Chem, 2013

R Badugu, J R Lakowicz and C D Geddes, Analyst, 2004

E. Stoye Chemistry World 2013

MULTIVERSE THEORY

PATEL V

ABSTRACT

Multiverse theory describes a hypothetical group of infinite and finite universes varying with different possibilities. These universes together make up all that exists and could potentially exist. In these other universes the fundamental constants, and even the basic laws of nature, may be varied. The theory that invokes these universes holds that such universes are popping into and out of existence and colliding all the time.

1. HYPOTHESIS

Two research papers published in *Physical Review Letters* and *Physical Review D* are the first to detail how to search for signatures of other universes. It's believed that disk-like patterns in the cosmic microwave background (CMB) radiation could provide significant evidence of collisions between other universes and our own. CMB is relic heat radiation left over from the Big Bang. Numerous modern theories of fundamental physics predict that our universe is contained inside a bubble. In addition to our bubble, this theoretical 'multiverse' will contain others, each of which can be thought of as containing another universe. In the other 'contained universes' the fundamental constants, and even the basic laws of nature, may be different. (UCL.ac.uk, 2011)

2. TEST

A team of cosmologists ran simulations of what the sky would look like with and without cosmic collisions. From this they put the first observational upper limit on how many bubble collision signatures there could be in the CMB sky. However Scientists, being human, tend to only see what they want to see. This phenomenon causes them to look at coincidences and judge them as an actuality. Dr Hiranya Peiris, a cosmologist at University College London, therefore said that data from the Planck telescope - a next-generation space telescope designed to study the CMB with greater sensitivity - would idea on a firmer footing, or even disprove it.

3. RESULT

George Efstathiou, director of the Kavli Institute of Cosmology at the University of Cambridge, noted that the theories that invoked the multiverse were fraught with problems, because they dealt in so many intangible or immeasurable quantities. "My own personal view is that it will need new physics to solve this problem but just because there are profound theory difficulties doesn't mean one shouldn't take the picture seriously." (BBC.co.uk, 2011)

4. CONCLUSION

If proven correct, the theory may hold the answers to questions of the universes' origin and maybe much more. However some peculiar, unexplained features of the results may well require new physics to be understood.

REFERENCES

UCL.ac.uk. (2011, August 3). *First observational test of the 'multiverse'*. Retrieved 2013, from UCL: http://www.ucl.ac.uk/news/news-articles/1108/110802-first-test-of-multiverse

BBC.co.uk. (2011, August 3). *'Multiverse' theory suggested by microwave background*. Retrieved 2013, from BBC News: http://www.bbc.co.uk/news/science-environment-14372387

AN INVESTIGATION INTO THE ENHANCEMENT OF FINGER-MARKS IN BLOOD ON FRUIT AND VEGETABLES

RICKMAN D

ABSTRACT

Studies have proved successful enhancement of latent fingerprints on fruit and vegetables. This study was made to establish the most effective technique of enhancing fingerprints in blood of various fruit and vegetables. The enhancement techniques used are protein stains (e.g. acid black 1) and amino acid stains (e.g. Ninhydrin). Different factors affect fingerprints such as ageing of fruit and vegetables used to assess the suitability and sensitivity of the various enhancement techniques. Different protein stains where most effective, for the enhancement of fingerprints in blood of fruit and vegetables. The aubergine and cucumber skins appeared to be the most accurate at enhancing fingerprints to different chemicals. On the other hand, no enhancement was shown in blood of the nectarine fruit.

1. HYPOTHESIS

Trapecar and Vinkovic compared two fingerprint powders and found that Swedish black powder was the best for getting ridge detail on different surface types. This implied black magnetic powder would be better in the fingerprinting of fruit and vegetables. (Matej Trapecar, 2008) The hypothesis of this study was to find out if magnetic fingerprint powder gives a better enhancement of fingerprints in blood on fruit and vegetables compared to common fingerprint powders. (Laura Rae, 2013)

2. TEST

Rae et al. bought locally sourced fruit and vegetables, aubergine, banana, cucumber, nectarines and oranges, to compare the colours and textures of different fruit and vegetables. They were kept within the lab for 1, 7, 14 and 21 days. One person used their fingerprints for all the tests, using horse blood. They were to place their finger within the blood and dab twice onto chemical free towels to remove any excess blood. 8 fingerprints were made on each fruit and vegetables from visible prints, to barely visible to latent. The different enhancement techniques were used on each fruit and vegetables and the visibility of the fingerprints was measured using the grading scale directly after and 24 hours after placing the fingerprint.

3. RESULT

The results from the study is that protein stains graded 3 and 4 fingerprints whereas Ninhydrin and DFO provided little enhancement.

4. CONCLUSION

Recovery of fingerprints is possible and non-magnetic enhancement techniques are better for the use of fruit and vegetables.

Bibliography

Laura Rae, D. G. (2013). An investigation into the enhancement of fingermarks in blood on fruit and vegetables. *Science and Justice*, 14.

Matej Trapecar, M. K. (2008). Techniques for fingerprint recovery on vegetable and fruit surfaces - A preliminary study. *Science and Justice*, 192-195.

DECYNIUM-22 AND FUTURE DEVELOPMENTS IN THE TREATMENT OF DEPRESSION USING SSRIS

RUDDY L.

ABSTRACT

The development of effective treatments for depression has been an important long term goal for both medical professionals and pharmaceutical companies alike, as depression can be an extremely debilitating mental illness for many sufferers. Existing drugs to treat depression include monoamine oxidase inhibitors (MAOIs) and tricyclic antidepressants (TCAs), but due to the side effects of these medications, the most commonly prescribed medications for the treatment of depression are SSRIs (selective serotonin reuptake inhibitors). Unfortunately, SSRIs are not effective for all patients, so Decynium-22 was synthesised with the possibility of the development of new drugs with different targets, to improve ineffective treatment.

1. HYPOTHESIS

SSRIs inhibit the serotonin transporter mechanisms (SERT) in presynaptic neurons in the brain, to attempt to increase the amount of extracellular serotonin in the synapses between neurons, which is associated with a greater rate of recovery from depression. However, it is now believed that SSRIs have proved ineffective in the treatment of some patients as there are other mechanisms responsible for the removal of serotonin from synapses between neurons in the brain, such as organic cation transporters (OTC1, OTC2 and OTC3) and plasma membrane monoamine transporters (PMAT) (Horton, et al., 2013). The hypothesis is that these mechanisms prevent the amount of serotonin associated with the effective recovery from depression from accumulating in synapses in the brain, despite the fact serotonin transporter mechanisms have been inhibited by SSRI medications.

2. TEST

Decynium-22 was synthesised to be tested in the brains of mice, to determine whether it inhibited the organic cation transporter mechanisms also thought to be responsible for the clearance of serotonin from synapses in the brain. The SSRI Fluvoxamine was administered to the mice as well as the new drug, Decynium-22 (Horton, et al., 2013).

3. RESULT

It was found that Decynium-22 increased the efficacy of Fluvoxamine in increasing the levels of extracellular serotonin found in the brains of the mice. This shows that Decynium-22 did block the organic cation transporter mechanisms responsible for the clearance of serotonin from synapses in the brains of mice.

4. CONCLUSION

As Decynium-22 has been shown to be effective in blocking mechanisms other than SERT, which clear serotonin from synapses in the brains of mice, similar drugs could be developed to inhibit these mechanisms in humans. This could lead to an improvement in the treatment of depression using SSRIs, as the new drugs could be used alongside SSRIs to improve their effectiveness in those who find that SSRIs alone do not fully treat the symptoms of their depression.

REFERENCES

Horton, R. E., Apple, D. M., Owens, A., Baganz, N. L., Cano, S., Mitchell, N. C., et al. (2013). Decynium-22 Enhances SSRI-Induced Antidepressant-Like Effects in Mice: Uncovering Novel Targets To Treat Depression. *The Journal of Neuroscience.*

THE IMAGING OF THE COVALENT BONDS IN A MOLECULE AS THEY ARE BROKEN AND FORMED BEFORE AND AFTER A REACTION
RUDDY L.

ABSTRACT

For the first time, the bonds between the atoms in a molecule have been observed as they are broken and formed during a reaction. Whilst it was once thought by many scientists to be impossible to actually see and photograph the bonds in a molecule, due to advances in atomic force microscopy this has now happened. The hydrocarbon $C_{26}H_{14}$ was observed as it underwent a series of cyclisation reactions to produce its structural isomers, and images of the structure of the reactant and products were produced using non-contact atomic force microscopy.

1. HYPOTHESIS

The cyclisation reactions of $C_{26}H_{14}$ were researched with the aim of manufacturing new nanostructures of graphene, a single layer of graphite. (Sanders, 2013) The images produced, showing single, double and triple covalent bonds were unexpected by the scientists involved.

2. TEST

The research was conducted at Berkeley lab in the US. The reactant was placed on a surface of Ag (100), which was then heated to temperatures above $90°C$ up to $150°C$ to catalyse the cyclisation reactions (de Oteyza, et al., 2013). The images were produced using the process of non-contact atomic force microscopy, in which a molecule of CO is adsorbed onto the atomic force microscope, leaving a single oxygen atom at the tip of the microscope. When the tip of the microscope is passed over a sample, the atomic force microscope measures the amount of repulsion between the electrons in the oxygen atom and the electrons in the sample molecule, creating an image.

3. RESULT

As there is a greater amount of repulsion between the oxygen atom and the sample molecule in the areas of the molecule with greater electron density, where there are single, double and triple covalent bonds, this allows the atomic force microscope to detect exactly where the bonds are (Giessibl, 2013)., producing photographs which resemble previous diagrams of the bonds in molecules.

4. CONCLUSION

As a result of being capable of observing the bonds in a molecule before and after a reaction, it is possible to develop controlled methods of manufacturing many different molecules, for example, graphene. This could lead to advancements in other fields, such as computing. These findings are also highly significant, as they prove that previously drawn diagrams of chemical bonds, and ideas of where and how they are formed are accurate.

REFERENCES

de Oteyza, D. G., Gorman, P., Chen, Y.-C., Wickenburg, S., Riss, A., Mowbray, D. J., et al. (2013). Direct Imaging of Covalent Bond Structure in Single-Molecule Chemical Reactions.

Giessibl, F. J. (2013). Seeing the Reaction. *Science*.

Sanders, R. (2013, May 30). Retrieved from newcenter.berkeley.edu: http://newscenter.berkeley.edu/2013/05/30/scientists-capture-first-images-of-molecules-before-and-after-reaction/

PERCEPTUAL CONVERGENCE OF LARGE MIXTURES IN OLFACTION IMPLIES AN OLFACTORY WHITE

SCAWIN J.

ABSTRACT

The idea behind white smell is that you can combine many different pure molecules of smell and these cancel each other out. White is a combination of signals at equal intensity across perceptual space. This also relates to how smell is detected.

1. HYPOTHESIS

The scientists wanted to look at how scents were detected. They also wanted to see if they could create white smell and how many scents it would take for them reach white smell.

2. TEST

They had 86 different pure scents, made up from single type of odour molecule. They then used these to create a smell map. There were 24 participants, who did not know the aims of the research. They then selected different scents from the smell map and combined these together. They asked the participant to describe the intensity of the scent. This was to make sure the scents they used were of equal intensity, because if they were different they could be thought to smell different because this changes the way it is detected. These final scents were the ones they used in the rest of the research. They were also asked to rate the similarity of two scents from one to nine, nine being the most similar. In another study with 20 participants, they had three jars, two with the same combination of 30 scents and one with a different combination but thought to be similar scent. They then had to rate their confidence on how similar they were to each other, from one to five, five being the most confident.

3. RESULT

They found white smell after mixing 30 different pure elements. In the study looking at the similarity of scents, they found that 147 comparisons were made to non-overlapping scents and 44 comparisons were made to overlapping scents. In the study with the 3 jars, they found that by discarding confidence level ratings who answered correctly in less than 70% of the trials and 16/20 results were kept.

4. CONCLUSION

The study of the comparison of scents to see how similar they were, suggest that scents thought to be non-overlapping were in fact thought to be similar, which suggest that scents are detected differently than first thought. They know from this research now that scents are detected as a whole scent not as separate scents than first thought. The findings from the 3 jar study suggests that most people can detect when 2 scents are similar and when they are different. White smell could be made by mixing over 30 pure scents, this does not have to be any particular scents in order, and it can be any mixture of pure scents as long as there are over 30 of them. This information can be used in the study of neurobiology.

REFERENCES

Wises et al-www.pnas.org

www.rdmag.com

November 19th 2012- WEISS T. SNITZ K. YABLONKA A. KHAN R M. GAFSOU D. SCHNEIDMAN E. SOBEL N.

MYOELECTRIC PROSTHETICS AND NEURO-REHABITALISATION

SINGADIA S.

ABSTRACT

Prosthetic limbs have been developed to imitate mundane realism, and now imitate functional qualities of an actual limb. Myoelectric prosthetics are controlled neurologically, through an interface. However, research shows that the EMG's can be distorted by changes in pH and movement on the surface in the skin. Electrical impulses can be transmitted faster and amplified further when fused with titanium to bone tissue of the person limb.

1. HYPOTHESIS

This process of neurological control of a prosthetic starts with Target Muscle Reinnervation (TMR), which is transferral of muscles from the amputated limb site to alternative nerve sight, which then transmit the EMG (electromyogram) signals to the surface of the skin to be processed by electrodes. The strength of the electrical impulses and information being transmitted and received by the prosthetic limb to move neurologically was found successful. Research into creating an interface in which information can be passed onto an external object was carried out (Nicolelis, 2013), and the result was to fuse electrodes onto residual bone tissue of the limb with titanium, for the weak EMG's to be amplified and processed into the prosthetic limb, the electrodes acting as an interface (Max Ortiz Catalan, 2012). Surgically implementing sensors into the stump of an amputee is a way to connect the human nervous system to the microprocessors in the prosthetic (Doshi, 2011).

2. TEST

For electrodes to be permanently implanted into the nerves and muscles, an operation is required, the electrodes anchored into the bone with a titanium screw, for more reliable control over the prosthetic (Branemark, 2013). Simulations were used for amputees to regain control over their limbs, practise movement in virtual-reality environment and decode algorithms. A monkey was used to test the artificial limb with a 1-D joy stick to achieve feeding himself in 4 basic movements.

3. RESULTS

After using the simulator, patients were able to control the artificial limbs in ways that they didn't know they could. Artificial limbs were permanently fused onto the stub of the missing limb, and the time taken for signals to be transmitted from the brain to the limb was closely achieved to the prosthetic. Laboratory experiments have so far been taken, but first operations for the public will start in the winter of 2013.

4. CONCLUSION

Prosthetics are developed enough to imitate actual amputated limbs, and can be neurologically controlled by the prosthetic being surgically attached to the stump of the limb, electrical signals received and processed by the motors of the prosthetic.

REFERENCES

Branemark, D. (2013) http://neurogadget.com/2013/04/08/osseointegrated-prosthetic-arm-controlled-via-direct-nerve-implants/7653

Catalan, PhD (2012) Thought-controlled prosthesis is changing the lives of amputees

R.A., Hwang, E.J., and Mulliken, G.H. (2010). *Cognitive Neural Prosthetic*

ARTIFICIAL PHOTOSYNETHESIS OF SUSTAINABLE ENERGY RESOURCES

SINGADIA S.

ABSTRACT

Sustainable energy resources are hard to come by, and are always developing to find the source which is most economically beneficial and provides the huge quantities of clean energy, without contributing to global warming. Our standards of living require more energy, and the rate will continue increase in the future, therefore a sustainable energy source is in immediate demand, in target countries such as India, Africa and China. Daniel Nocera was the inverter of the artificial leaf, an object composed of Silicon, Nickel and Cobalt, the system of photosynthesis built by placing tiny solar panels on microbes of the leaf, to maximise the conversion of visible light from sunlight, instead of ultraviolet light to methanol and other useful fuels, and oxidising water into oxygen and hydrogen gas by a platinum catalyst. The captured hydrogen gas would then be used in fuel cells to make electricity.

1. HYPOTHESIS

Light-driven water oxidation and hydrogen production by molecular catalysts are the stages of artificial photosynthesis in Daniel G. Nocera's artificial leaf. The use of two renewable resources, water and sunlight, was the start of the idea to take advantage of photosynthesis, as it converts the majority of the Earths most renewable source (the sun) into energy. However, leaves do not reach their full conversion rate, and the aim of the artificial leaf was to maximise the conversion of visible light into energy, originally in the form of hydrogen released from the oxidation of water. This hydrogen was then planned to be captured and used either directly into fuel cells as electricity, or used as a reactant to be processed as other fuels. An advantage of this technique which benefits the environment which other technologies have failed to do (burning of fossil fuels) is that the process produces energy without exhausting greenhouse gases.

2. TEST

Testing energy conversion methods on the micro-scale before applying them to large-scale systems was carried out by fully integrating microfluidic test-bed for assessing and optimizing solar-driven electrochemical energy conversion systems.

3. RESULT

The rate of natural photosynthesis was found to be slow, and not up to its full potential, at a rate of 100 to 400 turnovers per second. The artificial leaf was designed to reach a rate of 300 turnovers per second, resulting in a faster rate of photosynthesis overall.

4. CONCLUSION

There is a higher conversion rate of solar energy to fuel energy in the form of protons and electrons, with 5-10% conversion rate of solar energy. This energy can be used in electricity from fuel cells, and can power houses in 3rd world countries, from 1 bottle of water up to a day long.

REFERENCES

"Artificial photosynthesis breakthrough: Fast molecular catalyzer." ScienceDaily, 12 Apr. 2012. Web. 4 Jun. 2012.

American Chemical Society. "Secrets of the first practical artificial leaf." ScienceDaily, 9 May 2012. Web. 2 Jun. 2012.

ERIS: THE DWARF PLANET

STEELE J

ABSTRACT

Scientists have found a dwarf planet on the edge of our solar system, it is just past Pluto, which is called Eris. This dwarf planet is the 9th largest mass in our solar system. Eris was discovered on January 5th 2005, from photos that were taken on October 21st 2003. (Jr., 2005-10-16)

1. HYPOTHESIS

The hypothesis was that a team from Palomar observatory thought that there would be other massive masses past the last planet at that point, Pluto.

2. TEST

The way that they found this planet was by analysed the photos that they took in 2003. They found the planet by looking at the background stars and as the planet went across the background star the star disappeared.

3. RESULT

The result that has been from this test is that there is a dwarf planet on the edge of our solar system, therefore they called this planet Eris after a year of the dwarf planet not having an official name. Also by finding Eris they were able to create a new planet category, which is a dwarf plant, therefore this meant that they also categorised Pluto as a dwarf planet.

4. CONCLUSION

To conclude the discovery of this dwarf planet was important as scientists are gaining more information about our solar system and the universe.

REFERENCES

Beatty, K. (2010, November). *Former 'tenth planet' may be smaller than Pluto*. Retrieved from NewScientist.com.

Brown, M. (2007). *Lowell Lectures in Astronomy*.

Jr., T. H. (2005-10-16). *His Stellar Discovery Is Eclipsed*.

BIOCRIMINOLOGY - BRAIN ABNORMALITIES IN MURDERERS INDICATED BY BRAIN SCANS

JAHNA STIRK

ABSTRACT

A.Raine, looked at a group of participants who were charged with murder but pleaded not guilty due to insanity (NGRI). He was investigating the theory that there are 14 sections of the brain and the prefrontal cortex with are involved with violence. These sections of the brain have been known to be linked to unusual emotional responses such as lack of fear.

1. HYPOTHESIS

The aim of the experiment completed by A.Raine was to look at direct measures of both cortical and subcortical brain functioning using PET scans in a group of murderers who have pleaded NGRI. The expectation was that the murderers would show evidence of brain dysfunction in their prefrontal cortex as well as in other areas that are thought to be linked to violent behaviour (1).

2. TEST

The study used PET scans to examine the brains of 41 criminals (39 males and 2 female) that were charged with murder and were pleading NGRI, and 41 control participants who had been diagnosed with similar mental health issues to those claimed by the criminals but no history of murder(1). All the participants were kept medication free. All the participants were injected with a radioactive glucose tracer, and then given a PET scan. The NGRIs were compared with the controls on the level of activity in the left and right hemispheres of the brain in 14 selected areas which are thought to be linked to violence(2).

3. RESULT

Compared to the controls, the NGRIs were found to have less activity in their prefrontal and parietal areas, more activity in their occipital areas and no difference in their temporal areas. The results from the subcortical areas found less activity in the corpus callous, they also found an imbalance of activity in the hippocampus, compared to the controls, and the NGRIs had less activity in the left side and more activity in the right side. Also in the thalamus the NGRIs had more activity in the right side, although no difference in the left.(2)

4. CONCLUSION

Raine argues that the difference in activity in the prefrontal cortex can be seen to support theories of violence that suggest it is due to unusual emotional responses such as lack of fear(1). The results also show that they have inappropriate emotional expression.

REFERENCES

(Rainestudy.htm) Raine(1), A, Buchsbaum, M & LaCasse, L. (1997) Brain abnormalities in murderers indicated by positron emission tomography. Biological Psychiatry, 42, 495 - 508 (2)

ANTHROPOGENIC AEROSOL FORCING OF ATLANTIC TROPICAL STORMS

STRETTON M.

ABSTRACT

Since 186, there have been varying numbers of hurricanes each decade, with an increasing number over the past few years. These numbers could be forced by sea surface temperatures, but the amount of anthropogenic aerosols in the atmosphere also fit the pattern if they forced hurricanes, as when the concentration of aerosols increases, the number of tropical storms decreases. This happens because the brightness and the lifetime of low-level clouds increases (New Scientist, 2013), so the sea cannot gain as much heat from the sun and there is not as much energy to power the tropical storms.

1. HYPOTHESIS

The concentration of anthropogenic aerosols in the atmosphere affects the frequency of tropical storms.

2. TEST

A range of climate simulations were created, removing one forcing factor at a time, and then calculated the contribution of each factor on the frequency of tropical storms. The factors that were tested were greenhouse gases, sulphate aerosols, wind shear, and precipitation in the main hurricane development region. (Dunstone, et al., 2013)

The factor that had the highest contribution was the concentration of anthropogenic aerosols. New simulations were created. They altered the concentration of the anthropogenic aerosols in the atmosphere to ones that match the concentration found in decades from 1860- present day. The numbers of hurricanes in the simulation and from real life data were then compared.

3. RESULT

Changing the concentration of anthropogenic aerosols created a pattern of hurricanes that matched the real life data. These results show that as the concentration of aerosols in the atmosphere decreases, the number of hurricanes increases. This is because the aerosols change the brightness and lifetime of clouds, and increases the wind shear in the Atlantic Ocean, giving less energy to create tropical storms and destroying new tropical storms before they reach the mainland. However, the concentration of greenhouse gases also increased and decreased in the same pattern as the anthropogenic aerosols. They were not seen to of cause the hurricanes and tropical storms in the past though, as they have a lifetime of 80 years. This will mean that the effects of the greenhouse gas concentration in the atmosphere will be seen after the next two decades, or after the effects of the aerosols.

4. CONCLUSION

For the next two decades, if the pollution controls are tightened further, then there will be an increase in tropical storms.

REFERENCES

Want fewer hurricanes? Pollute the air. (2013, June 23). Retrieved from New Scientist: http://www.newscientist.com/article/dn23743-want-fewer-hurricanes-pollute-the-air.html#.UdP83PmmhtM

Dunstone, N. J., Smith, D. M., Booth, B. B., Hermanson, L., & Eade, R. (2013). *Anthropogenic aerosol forcing of Atlantic tropical storms.* N J Dunstone.

DETECTING TAU NEUTRINOS FROM THE OSCILLATION OF MUON NEUTRINOS

STRETTON M.

ABSTRACT

Scientists have observed that when neutrinos are emitted from stars like the sun, a smaller number of neutrinos than the star emits reach the surface of the Earth (OPERA, 2013). Neutrino research will aim to explain the reason behind this, and experiments such as CNGS are behind the research. Beams of muon neutrinos and muons are created by colliding protons with each other. The products of these collisions are then accelerated from CERN in Geneva, 730 kilometres to Laboratori Nazionali del Gran Sasso in Italy. The oscillations of the muon neutrinos cause the neutrinos themselves to change flavour to tau neutrinos.

1. HYPOTHESIS

When beams of muon neutrinos are accelerated, some will change flavour to tau neutrinos.

2. TEST

Protons were collided with each other and focussed into a beam using a graphite target. The products of the collision were kaons and pions, which are very unstable. These particles will then decay to form the muons and muon neutrinos. (OPERA,)

This beam was then accelerated from CERN in Geneva, Switzerland to LGNS in Gran Sasso, Italy. This journey is 730 kilometres long and takes the beam three milliseconds (OPERA,). In the Italian laboratory, their detector OPERA collected the results.

3. RESULT

There have been successful results found on three occasions; in 2010, 2012 and 2013. They all showed that tau neutrinos were discovered from the beam of muon neutrinos by the OPERA detector, and therefore proved the hypothesis to be correct. However, in the first test, the beam arrived 60 nanoseconds earlier than expected, which would have meant that the neutrinos had travelled faster than the speed of light. This was later found out to be because of a flawed fibre-optic cable, and therefore the results were in fact correct and neutrinos do not travel that fast. This was then confirmed by the other two discoveries that arrived at the time expected.

In order for neutrinos to oscillate, they must have mass, and therefore this experiment has confirmed that neutrinos have mass. This means there will be further research that is going to be conducted, such as Fermilab to MINOS.

4. CONCLUSION

Neutrino oscillations as a result of acceleration do cause the neutrinos to change flavour. This is demonstrated by muon neutrinos changing to tau neutrinos in the Cern Neutrinos to Gran Sasso experiment. It also confirmed that neutrinos have mass.

REFERENCES

OPERA. (2013, May 26). *OPERA*. Retrieved from New neutrino oscillation event discovered at OPERA: http://operaweb.lngs.infn.it/spip.php?article58

OPERA. (n.d.). *OPERA*. Retrieved from The neutrino beam: http://operaweb.lngs.infn.it/spip.php?rubrique41

OPERA. (n.d.). *OPERA*. Retrieved from About OPERA: http://operaweb.lngs.infn.it/spip.php?rubrique1

TEEN GIRLS WHO EXERCISE ARE LESS LIKELY TO BE VIOLENT

STUART J.

ABSTRACT

Exercise is a common topic of investigation, resulting in various links between affecting psychological issues (such as reducing depression and increasing self-esteem) and exercising regularly being found. A recent study in which 1,312 students in an inner city area of the USA has uncovered another possible positive side-effect of exercise: that regular exercise in adolescent girls can reduce their involvement in violent behaviour (al R. N., 2013). Despite further research being needed to confirm this relationship, even the possibility of exercising making the streets safer is a major breakthrough and could be used to make a difference to communities worldwide.

1. HYPOTHESIS

Exercise has an effect on adolescent exposure to violence.

2. TEST

The investigation was a secondary analysis of a previously conducted 'Partners and Peers' study of 4 high schools in New York City (al, 2008). Questions on the study were taken from the Youth Risk Behaviour Survey and the Child Health Illness Profile- Adolescent Edition which was conducted in 2007/8. A variety of students completed the survey: 56% were female, 73 % were Latino and 19% were Black. In the study, exercising regularly was measured by four variables. These variables were: exercise frequency in the past four weeks (less than 10 days or greater than 10 days); number of sit-ups in the past four weeks (less than 20 or more than 20); longest run in the past four weeks (less than 20 minutes or more than 20 minutes) and playing in an organised team in the past 12 months (0 months, 1-2 months or 3+months). There were four possible outcomes to represent the exposure to violence of each teenager. These four outcomes were: carrying a weapon in the past 30 days (0 days or more than one day); being in a physical fight in the past 12 months (0 times or more than once); being in a gang in the last 12 months (yes or no). The results were also stratified by gender because of an interaction. (Teen Girls Who Exercise Are Less Likely to Be Violent, 2013)

3. RESULT

For females, all four of the possible variables resulted in a decreased violence outcome. For males, the only significant variable was being in a team which caused a reduced likelihood of being in a fight.

4. CONCLUSION

There is a clear link between regular exercise and decreased exposure to violence for females. However, this link is not seen as obviously for males.

REFERENCES

Teen Girls Who Exercise Are Less Likely to Be Violent. (2013, May 6). Retrieved from Science Daily: http://www.sciencedaily.com/releases/2013/05/130506095405.htm

al., D. A. (2008). Partners and Peers. *Sexual and Dating Violence Among NYC Youth*.

Romo, N. D. (2013). The Pediatric Academic Societies. *The Effect of Regular Exercise on Exposure to Violence in Inner City Youth*.

CONSUMPTION OF TRANS FATS CAUSE AGGRESSION

STUART J.

ABSTRACT

Dietary trans fatty acids (dTFA) are usually the products of hydrogenation and are found in many foods such as margarines, shortenings and prepared foods. As it is known that dTFA inhibit the production of omega-3 fatty acids (which have been experimentally proven to reduce aggression), it was decided that it would be beneficial to investigate the side effects of dTFA. Adults of both genders were questioned about their eating habits so nutritional data could be gained and compared against different endpoints that measured the aggression of the individual. As was expected, the results of the investigation provided evidence supporting the idea that dTFA may cause aggression (al, 2013).

1. HYPOTHESIS

It was theorised that dTFA may be a cause of greater aggression and irritability.

2. TEST

1018 adults (both male and female) who were at least 20 years old, were examined so that they could take part in a lipid lowering therapy called the UCSD Statin Study. 945 of these had completed a dietary assessment before their visit and these people were targeted for this present study. A variety of people were studied apart from those who were on lipid medications, had low/high LDL levels or those who had diabetes/cardiovascular disease/HIV/cancer. Data regarding the subject's nutrients were collected using a food frequency questionnaire developed by the Nutrition Assessment Shared Resource of the Fred Hutchinson Cancer Research Centre. These were compared against possible behavioural endpoints aiming to measure the subjects levels of aggression. These endpoints were: Overt Aggression Scale Modified (OASMa), Life history of aggression, Conflict Tactics Scale, Impatience and Irritability. Both Impatience and Irritability were a self-report of subjective Impatience/ Irritability that asked the subjects to rate each factor on a scale of 1-10 (1 being not present, 10 being maximally present). The results were adjusted to take into account possible reported aggression predictors like age, sex, education, alcohol, smoking and exercise. (Science Daily, 2013)

3. RESULT

Analysis of the results showed a significant association between dTFA and each aggression-related endpoint. The association remained after the results were adjusted to take into account the identified aggression predictors.

4. CONCLUSION

dTFA showed a strong link to behaviours that have negative effects on others. In fact, dTFA were more predictive of this negative behaviour than other assessed and acknowledged aggression predictors.

REFERENCES

(2013, May 6). Retrieved from Science Daily: http://www.sciencedaily.com/releases/2013/05/130506095405.htm

al, G. B. (2012). Trans Fat Consumption and Aggression. *PLos one*, 1-5.

NEW TYPE OF FRICTION

STUART J.

ABSTRACT

In all natural systems, friction is present; it is the cause of both wear and energy loss in various places including machines and joints. Many scientists are searching for methods in which friction can be minimised in order to make processes more efficient by reducing energy loss. During the search for low-friction components, a team of scientists from Technische Universitaet Muenchen have found another type of friction (the others being sliding, rolling and static) that the physicists have names 'desorption stick' (Balzar, 2013).

1. HYPOTHESIS

Single polymer molecules in various solvents slide over or stick to surfaces differently.

2. TEST

The scientists used single polymer molecules to conduct their experiment. They joined the end of one molecule to the nanometer-fine tip of a very powerful atomic force microscope. During the time in which they pulled the molecule over different surfaces, the microscope measured the resulting forces. From these detected forces, the researchers could deduce how the polymer coil behaved. (Science Daily, 2013)

3. RESULT

The physicists detected a friction mechanism for some combinations of polymer, solvent and surface other than the expected mechanisms. The polymer does stick to the surface but it can also be pulled from its coiled conformation into the solution around it without a significant force being exerted. (Friction in the nano-world, 2013) It was assumed by physicist Thorsten Hugel that the cause is probably a very low internal friction within the coil of the polymer. This internal friction is what has been called 'desorption stick'.

Also, it was found that desorption stick does not depend on the speed of the movement of the molecule or the support surface or the adhesive strength of the polymer. Actually, the desorption stick depends in the chemical nature of the surface and the quality of the solvent.

4. CONCLUSION

There are not only the types of friction as were previously thought, instead there is a new type that has been named 'desorption stick'. This desorption stick depends on the solvent that the substance is immersed in and the nature of the surface.

REFERENCES

Friction in the nano-world. (2013, May 15). Retrieved from TUM website: http://www.tum.de/en/about-tum/news/press-releases/short/article/30843/

Friction in the Nano-World: Physicists Discover a New Kind of Friction. (2013, June 17). Retrieved from Science Daily: http://www.sciencedaily.com/releases/2013/05/130515113831.htm

Balzar, B. e. (2013). Nanoscale Friction Mechanisms at Solid–Liquid Interfaces. *Angewandte Chemie International Edition.*

COCAINE ADDICTION

VINCENT. L

ABSTRACT

For my presentation I chose to research all about the drug cocaine and its effects on the body. I'm going to look at what cocaine actually is, what chemicals might be found in the drug and where it comes from. I'm going to look at what the substance does to the body, how it makes you feel and the long and short term effects. I am also looking at some research about cocaine addiction and how it can be controlled.

1. HYPOTHESIS

The piece of research that I have looked at was by the National Institute of Drug Abuse. It looked at addiction of cocaine in drug addicts and how they are able to inhibit certain parts of the brain that are to do with drug addiction. By inhibiting parts of their brain they are able to gain control of their cravings and can reduce them so that when they watch a video of someone using cocaine, there craving is drastically lower than someone who is not inhibiting their brain.

2. TEST

To test their hypothesis, the National Institute of Drug Testing used 24 cocaine addicts who had to watch a short video of someone using cocaine. Half of the group were told to inhibit the parts of the brain that control addiction whereas the other half were not told to do anything. They used PET scans and radioactive glucose to compare the brain activity of the two groups.

3. RESULT

The results of the research show that cocaine addicts are able to reduce their addiction by gaining control of the areas of the brain associated with the addiction.

4. CONCLUSION

The research found that cocaine addicts, who were told to inhibit the parts of the brain that control addiction, felt their craving less when watching a video of cocaine use.

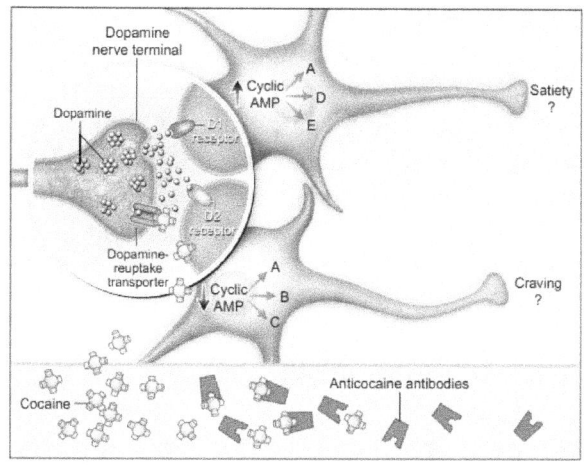

REFERENCES

http://www.stopcocaineaddiction.com/Where-does-cocaine-come-from.htm

http://www.talktofrank.com/drug/cocaine

http://www.addictionsearch.com/treatment_articles/article/risks-associated-with-cocaine-use-increase-due-to-levamisole_152.html

The National Institute of Drug Abuse-
http://www.sciencedirect.com/science/article/pii/S1053811909011604

THE MEANING OF DREAMS

EMILY WALE

ABSTRACT

In my presentation I will be looking at the meaning of dreams, looking at the psychology side, Freud's theory of what dreams mean and then looking at the recent research that science has done into looking into dreams, and then compare the science and the psychology side. I will talk about the study that has taken place into the study of dreams into science and their outcomes of the study. There is no definite answer to dreams yet as there is much more research that has to be taken place, all the things I will talk about are theories or research that has taken place, none are finalised answers.

1. HYPOTHESIS

The aim of this scientific experiment is to see if the same part of the brain is responsible for imagery processing, while you are asleep and also awake. A FMRI was used while they were awake and also asleep.

2. TEST

A neuroscientist (Yukiyasu Kamitani) and colleagues investigated at the ATR computational neuroscience laboratory in Kyoto, Japan. They used brain scans to look at different part of the brain that was used during sleep. They tested 3 adult men; they were awakened during the middle of them sleeping and then asked what images they remember seeing. During this they had an MRI, to see what areas of the brain were being used. They were awakened several times per hour. Then then split the different images they saw into 20 different categories. After the experiment they then showed the men images that contained the images from their dreams, they conducted a brain scan at the same time to see if the same part of the brain lit up to the same image.

3. RESULT

These dream reports shows that the volunteers dreamed in large part about ordinary things related to daily life with the occasional detour into the odd or fantastical. Brain activity during dreaming is similar to the brain activity associated with processing visual information in a waking state. The neural activity was so similar that researchers were able to predict the visual content of their dreams with 75 -80 per cent accuracy. However, decoding colour, action or emotion is also still beyond the scope of technology. From this they plan to have a follow up study into looking at REM sleep, the stage of most of our dreaming and most of our emotionally resonant dreams take place.

4. CONCLUSION

That dreaming does mean something to do with our everyday lives; however there is still so much we don't know yet about the purpose of sleep. Despite so much investigation into it there is still lot to solve.

REFERENCES

http://www.wired.com/wiredscience/2013/04/dream-decoder/

http://www.scientificamerican.com/article.cfm?id=the-science-behind-dreaming

https://en.wikipedia.org/wiki/Dream

http://www.huffingtonpost.com/dr-michael-j-breus/dream-meanings_b_2238609.html

http://www.dreammoods.com/dreaminformation/dreamtheory/freud3.htm

THE FORMATION OF LIFE ON TITAN

WARD AJG.

ABSTRACT

Titan is one of Saturn's moons. Saturn's giant moon Titan is the only known planetary-sized body – apart from the Earth – to sport a thick nitrogen rich atmosphere. A probe was sent out to titan, Cassini, which was able to take some samples of titan and discover if the conditions conductive to life. The results have enabled scientists develop specific instruments on future missions that could seek out these nitrogen-rich molecules.

1. HYPOTHESIS

An important step in determining models for the formation of life includes investigating nitrogen rich bodies such as Earth and Titan. To perform this experiment NASA would need to send out a probe to Saturn's moon titan. This experiment was supported by scientists at the University of Arizona who have performed laboratory experiments that show how atmospheric nitrogen can be incorporated into organic molecules. If there was a nitrogen layer on titan then there would be a possibility of the formation of life. Titan appears to have lakes of liquid ethane or liquid methane on its surface, as well as rivers and seas. For this reason some scientists believe that titan could contain some formation of life.

([i] Jia-Rui Cook, Cathy Weselby, NASA News release, "What is Consuming Hydrogen and Acetylene on Titan?", 2010)

2. TEST

Experiments were conducted by Hiroshi Imanaka and Mark Smith using the Advanced Light Source at Lawrence Berkeley National Laboratory. By irradiating a nitrogen-methane gas similar to the composition of Titan's atmosphere with high energy ultraviolet rays to simulate the effects of solar radiation on Titan's atmosphere. This experiment would indicate if the conditions necessary for the formation of life were possible.

3. RESULT

After analysing data from the Cassini–Huygens mission reported anomalies in the atmosphere near the surface which could be consistent with the presence of methane-producing organisms. Imanaka and Smith suspect that such compounds are formed in Titan's thick atmosphere and eventually fall to the moon's surface where they could be exposed to the conditions that enable the evolution of life.

4. CONCLUSION

The results of these experiments could help scientists develop specific instruments on future missions that could seek out these nitrogen-rich molecules.

REFERENCES

WARD AJG. Hadhazy, "Scientists Confirm Liquid Lake, Beach on Saturn's Moon Titan", Scientific American magazine, July 30, 2008 Jia-Rui Cook, Cathy Weselby, NASA News release.

AN INHIBITOR SHOWING EVIDENCE OF PREVENTING INFERTILITY IN YOUNG, FEMALE CISPLATIN CHEMOTHERAPY PATIENTS

WARD B.

ABSTRACT

A lot of young women who undergo Cisplatin chemotherapy for cancer related illnesses often become infertile due to the premature cell death of their egg cells, called oocytes. These are destroyed when the process of apoptosis is stimulated, causing early menopause. Cisplatin induces TAp63 and c-Abl kinase production, which allows the metabolic process of apoptosis to continue until cell death.

1. HYPOTHESIS

To prove that the inhibitor imatinib mesylate could stop the early onset destruction of juvenile eggs, called oocytes; by stopping the apoptosis before it initially occurs. Cisplatin encourages the production of the gene TAp63 which in turn encourages the production of the gene TAp73 which leads to apoptosis.

2. TEST

Ovaries from 5 day old mice (Paul, 2013) were extracted and cultured in test tubes with imatinib mesylate and Cisplatin. The test tubes were then left for approximately 96 hours before the ovaries are inserted into a kidney capsule and placed into mice. This was then left for a number of weeks to see if the process of apoptosis took place (Kim, et al., 2013).

3. RESULT

The results were not the expected results, as apoptosis had begun and DNA damage had already occurred to the oocytes. However, the apoptosis hadn't been completed as the inhibitor imatinib mesylate had inhibited c-Abl kinase which would originally produce a product to stabilise TAp63 to move onto the next stage of the process involving TAp73. The presence of imatinib mesylate means that only cell damage occurs and that fatality of the cell doesn't occur.

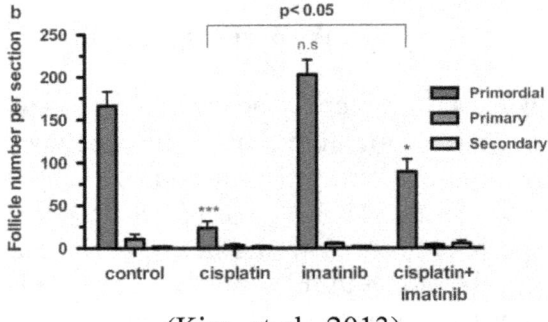

(Kim, et al., 2013)

4. CONCLUSION

The discovery is that the inhibitor imatinib mesylate blocks a part of the process of apoptosis which is caused by the presence of Cisplatin; giving the overall affect that the women who have undertaken Cisplatin chemotherapy will remain fertile after treatment.

REFERENCES

Kim, S.-Y., Cordiero, M., Serna, V., Ebbert, K., Butler, L., Sinha, S., et al. (2013). Rescue of platinum-damaged oocytes from processed cell death through inactivation of the p53 family signaling activity. *Cell Death and Differentiation.*

Paul, M. (2013, June 18). *Preventing Eggs' Death From Chemotherapy.* Retrieved June 26, 2013, from Northwestern University: http://www.northwestern.edu/newscenter/stories/2013/06/preventing-eggs-death-from-chemotherapy.html

3 LOW-MASS SUPER-EARTH'S FOUND IN HABITABLE ZONE OF THE STAR GLIESE 667C

WARD B.

ABSTRACT

Extrasolar planets and the possibility of life outside of the Earth has always been a fascination of astrophysicists. The star Gliese 667C has been found to have had 1 extra-solar low-mass planet in its habitable zone; however this has been tested against as it was thought that the previous results weren't analysed properly. With the collection of new data, from different technology, new evidence has been found to suggest there's more than one. (Three planets in habitable zone of nearby star (w/video), 2013)

1. HYPOTHESIS

To collect more data using Doppler spectroscopy, to compare and analyse the new and old results to conclude whether there is more than one Super-Earth. (Gilster, 2013)

2. TEST

To send Doppler signals from Earth to Gliese 667C and measures how the velocity of the star changes because another smaller object is orbiting it, for example a planet. The data would then be analysed, allowing scientists to work out the mass of the planet and the period of the planet. (Angala-Escude, et al., 2013)

3. RESULT

Several planets were found to be in the habitable zone of Gliese 667C instead of just 3 and further information showed that instead of 1 low-mass super-Earth in the habitable zone, 3 low-mass super-Earths happened to be there.

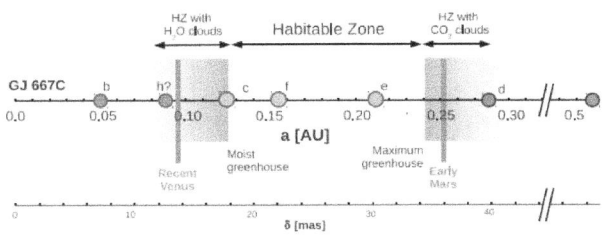

(Angala-Escude, et al., 2013)

4. CONCLUSION

There are several planets in the habitable zone of Gliese 667C and shows that in order to find extra-solar, habitable planets, you don't need to look at every nearby star to find many, just look at one and scientists may find many.

REFERENCES

Three planets in habitable zone of nearby star (w/video). (2013, June 18). Retrieved June 26, 2013, from Phys Org: www.phys.org/news/2013-06-planets-habitable-zone-nearby-star.html

Angala-Escude, G., Tuomi, M., Gerlach, E., Barnes, R., Heller, R., Jenkins, J., et al. (2013). A dynamically-packed planetary system around GJ 667C with three super-Earths in its habitable zone. *Astronomy & Astrophysics*, 1-3.

Gilster, P. (2013, June 25). *Gliese 667C: Three Habitable Zone Planets*. Retrieved July 1, 2013, from Centauri Dreams: http://www.centauri-dreams.org/?p=28177

List of delegates

Adams, Kyle

Aggarwal, Vipul

Attewell, Christian

Brown, Aidan

Brown, Hannah

Bruce, Cydney

Bruce, Olivia

Challenger, Benjamin

Coleman, Henry

Cooper, Daniel

Crossley, Leah

Deacon, James

Dexter, Miranda

Doe, Benjamin

Essor, Alysa

Goodband, Harry

Gulley, Charles

Hall, Sophie

Hartop, George

Hickenbotham, Demi

Irshad, Hanan

Katsaros, Alexandros

Lines, Omar

Lynch, Evie

McLeod, Katherine

Meachem, Benjamin

Naik, Nidhi

Pack, Anna-Marie

Pancholi, Pooja

Patel, Vishal

Rickman, Danielle

Ruddy, Lisa

Scawin, Jenny

Singadia, Sheena

Steele, Joseph

Stirk, Jahna

Stretton, Megan

Stuart, Jessie

Truong, Khang

Vincent, Lydia

Wale, Emily

Ward, Adam

Ward, Bethany

Whait, Henry

List of judges

Dr Boyce

Mr Nakeshree

Dr Meakin

Mrs Simpson

Photographs from the conference meal

www.ingramcontent.com/pod-product-compliance
Lightning Source LLC
Chambersburg PA
CBHW081051170526
45158CB00006B/1936